Introduction to Protein-DNA Interactions

Structure, Thermodynamics, and Bioinformatics

ALSO FROM COLD SPRING HARBOR LABORATORY PRESS

Other Titles of Interest

Bioinformatics: Sequence and Genome Analysis, Second Edition

Genes & Signals

A Genetic Switch, Third Edition: *Phage Lambda Revisited*

Molecular Cloning: A Laboratory Manual, Fourth Edition

Introduction to
Protein–DNA Interactions

Structure, Thermodynamics, and Bioinformatics

GARY D. STORMO, PH.D.

COLD SPRING HARBOR LABORATORY PRESS
Cold Spring Harbor, New York • www.cshlpress.org

Introduction to Protein–DNA Interactions
Structure, Thermodynamics, and Bioinformatics

Publisher	John Inglis
Acquisition Editors	Ann Boyle and Kaaren Janssen
Director of Editorial Development	Jan Argentine
Developmental Editor	Judy Cuddihy
Project Manager	Maryliz Dickerson
Permissions Coordinator	Carol Brown
Production Manager	Denise Weiss
Production Editor	Rena Steuer
Compositor	Techset Ltd.
Cover Designer	Ed Atkeson

Front cover: Computer-generated structural diagram showing the overall geometry of the Lac repressor protein binding to *lac* operator DNA (image generated using Pymol software from data in the Protein Data Bank database, entry 2KEI).

Library of Congress Cataloging-in-Publication Data

Stormo, Gary.
   Introduction to protein-DNA interactions : structure, thermodynamics, and bioinformatics / Gary D. Stormo.
      p. ; cm.
   Includes bibliographical references and index.
   ISBN 978-1-936113-49-1 (hard cover : alk. paper) – ISBN 978-1-936113-50-7 (pbk. : alk. paper)
   I. Title.
   [DNLM: 1. DNA-Binding Proteins–pharmacokinetics. 2. Binding Sites. 3. DNA–chemistry. 4. Protein Binding. 5. Transcription Factors. QU 58.5]

572.8'6459–dc23
                    2012035448

To my parents, Milo and Claryce, who gave me the love
of learning and the encouragement to follow wherever that led.

To my wife, Susan Dutcher, and my children, Ben and Adrienne,
who have enriched my life immeasurably.

Contents

Preface

THE BIOLOGICAL IMPORTANCE of PROTEIN–DNA INTERACTIONS has been recognized since the early 1960s, starting with the discovery by Jacob and Monod of the *lac* operon and its regulation in *Escherichia coli*. In the intervening 50 years, studies of protein–DNA interactions have made significant contributions to most areas of molecular, cellular, and developmental biology. A wide range of approaches has been applied in those studies, but they can be broadly classified into the three types that are the focus of this book: structural, thermodynamic, and bioinformatic. The earliest studies used biochemical and biophysical methods to analyze the thermodynamic and kinetic aspects of protein–DNA interactions. The first binding site sequences were determined in the early 1970s, which led to hypotheses about recognition mechanisms and the information required for regulatory systems to function. Technological advances in the late 1970s and the early 1980s, including the ability to sequence and synthesize DNA and to clone, express, and purify large quantities of proteins, facilitated many new types of studies. The earliest bioinformatics approaches were developed in the late 1970s, as soon as there were enough sequences for statistical analyses to be worthwhile. Shortly after that, as it became much easier to synthesize and purify sufficient quantities of specific proteins and DNA sequences of interest, structural studies rapidly increased. Further technological advances in the last two decades have continued to accelerate the pace of discovery. Most important have been further efficiencies in DNA sequencing that have resulted not only in whole-genome sequences for many species but also whole-genome and mRNA sequences from individuals as well as a variety of other sequence-based data sets. Our understanding of protein–DNA interactions and their roles in a wide range of biological processes has grown enormously, but there is still much we do not know and the field continues to be ripe for further discovery.

The primary goal of this book is to provide an introduction to protein–DNA interactions that bridges the three classes of approaches. Experts in any of the fields are not likely to learn anything new within their field; in fact, they will undoubtedly find examples of details being glossed over in favor of a simplified presentation. But experts in one area tend to have more cursory knowledge of the other fields and thus may learn from other sections of the book. Those who are new to the study of protein–DNA interactions or those outside the field with a casual interest in the topic may gain new insights throughout the book. If so, the book has succeeded even beyond the fact that I learned something in the process of writing every chapter.

The regulation of gene expression has fascinated me since my graduate school days. I have ventured into other topics, mostly related to how computer programs can help to uncover biological knowledge, but the majority of my efforts have been focused on understanding how networks of transcription factors regulate gene expression and control cell fates and phenotypes. I have been extremely fortunate to have been associated throughout my career with teachers and students, colleagues and collaborators, and most of all friends who have taught and encouraged me and made my whole adventure enjoyable. The list of those who made significant contributions to my research, many of whom I have never met but have benefited from immensely through reading their papers, is too long to include in this preface. But a few have had such a large influence that I must thank them here. Larry Gold, my graduate and postdoc advisor, kept research always fun and gave me the freedom and encouragement to follow an unconventional path. Tom Schneider, a fellow student in Larry's lab, and Andrej Ehrenfeucht, a mentor in all things computational, were there from the beginning and opened my eyes to new horizons that I would have missed without them. I have had many great collaborators over the years but special thanks go to John Heumann, Alan Lapedes, and Charles "Chip" Lawrence, each of whom has filled gaps in my knowledge and provided numerous insights into my own work that were initially invisible to me. I have also had many great students and postdocs who made progress possible and who taught me at least as much as I taught them.

This book would not have happened with the support and encouragement of the individuals at Cold Spring Harbor Laboratory Press, including Ann Boyle, Maryliz Dickerson, Kaaren Janssen, and Rena Steuer. Judy Cuddihy, in particular, made numerous improvements and helped at every step. I also thank those authors and publishers who allowed me to use their figures.

CHAPTER 1

Importance of Protein–DNA Interactions

THE WORD PROTEIN AND THE ACRONYM DNA HAVE BECOME staples of the English language, and the "central dogma of molecular biology," paraphrased as "DNA makes RNA makes protein," is familiar to anyone with even a moderate exposure to modern biology. The central dogma conveys the idea that DNA is the genetic material, stably passed down through the generations and containing the instructions for making all of the cell's proteins. Proteins embody the functional components of the cell, commissioned to perform the activities of living cells. A common and apt analogy is that DNA is the blueprint that specifies what the cellular factory makes, and the proteins are the products made by the factory. Of course, the entire machinery of the factory is made within the factory itself, making it self-contained and self-perpetuating—its only requirements are sources of energy and raw materials such as carbon, nitrogen, oxygen, and other essential elements. Although certainly an oversimplification, this idea captures the basic division of labor between the information-carrying DNA and the operative proteins. But left out of that basic description of the flow of biological information are the critical functions that the proteins perform on DNA. By itself, DNA is quite inert, chemically stable for long periods of time but utterly inactive. Proteins are required not only for the replication of DNA into the copies passed on to each new cell but also for all of the processes involved in interpreting the DNA information and synthesizing all of the machinery of the cell, including the proteins themselves.

Many proteins perform functions on DNA. A large collection of proteins are involved in replication, the DNA polymerase is responsible for synthesizing the new DNA strands, and an assortment of accessory proteins facilitate this process and increase its fidelity. When DNA is damaged, a variety of proteins are mobilized to effect its repair. Other proteins can create modifications to the DNA, such as methylases that add methyl groups to some of the bases and those that can recombine chromosomes, which exchange segments between homologous pairs and increase the genetic diversity of the resulting chromosomes. Proteins are also required for compaction and organization of chromosomes and their carefully controlled separation into daughter cells at the close of replication.

This volume is about one class of proteins that perform their functions on DNA—the sequence-specific DNA-binding proteins that control the transcription of DNA into RNA, the intermediate step between DNA and protein synthesis. Transcription is a complicated process involving a large collection of proteins. RNA polymerase performs the actual process of synthesizing RNA as a faithful copy of the DNA sequence. Its work is facilitated by many accessory proteins, or cofactors, that are required for the initiation, elongation, and termination phases of transcription. Where and when transcription of a particular gene occurs is governed by sequence-specific DNA-binding proteins, referred to throughout the book as transcription factors (TFs), that can function to turn on or off the expression of specific genes. TFs recognize and bind to specific DNA sequences so that their regulatory effects are confined to the genes with the appropriate sequences nearby rather than affecting the expression of all genes. This class of proteins constitutes a sizable fraction of the gene repertoire in most species, typically about 5% of the total number of genes in species as diverse as bacteria, yeast, worms, flies, plants, and mammals.

As with most important scientific discoveries, the determination of the structure of DNA served as both the end of one search and the emergence of whole new areas of research. DNA's double-helical structure and complementary base pairing raised new questions that were not even imagined until its structure was known. For example, the DNA structure led to experiments showing that genes are linearly contiguous stretches of DNA sequence and encoded digitally, therefore requiring a code to convert the four-letter DNA alphabet into the 20-letter protein alphabet. The nature of the code generated much theoretical speculation but was worked out experimentally in about a decade. It was also discovered that the information in the DNA was not translated directly into proteins—there was an intermediate molecule, messenger RNA (mRNA). The molecular machinery to synthesize proteins was discovered to be the ribosome, a particle identified in electron micrographs and named for its ribonucleoprotein (RNA and protein) content, but whose function was previously unknown.

One important question derived from the structure of DNA was how gene expression could vary among different cells if they have the same gene content. Different types of cells in our bodies, such as muscles, nerves, liver, and blood, have different functions and contain different sets of cell-type-specific proteins. If each cell contains the same DNA, there must be mechanisms to turn specific genes on and off so that each cell type synthesizes the proteins needed for its particular functions. This is the mystery posed by development—how a single-cell zygote can give rise to a multicellular organism with all of the appropriate cell types in their appropriate proportions, each containing the same DNA but expressing only a subset of the complete gene repertoire. Furthermore, the same cell can express different genes depending on its local environment. As with many aspects of biology, the first clues about the processes of regulating gene expression were discovered from studies on bacteria and simple eukaryotes. Although the examples from higher eukaryotes are much more complicated, the basic principles of gene regulation, mediated by sequence-specific TFs, can be illustrated with two classic examples—the *Escherichia coli lac* operon and the bacteriophage λ genetic switch.

THE *lac* OPERON OF *E. coli*

The bacterium *E. coli* will grow on many different sugars, including glucose, a simple monosaccharide, and lactose, a disaccharide composed of glucose and galactose. To grow on lactose, the *E. coli* cells must make the enzyme β-galactosidase, which cleaves lactose into the two monosaccharides. If *E. coli* cells are growing rapidly on glucose and the growth medium is switched to lactose, there is a lag period before rapid growth resumes. If *E. coli* cells are given both glucose and lactose, they first consume the preferred sugar, glucose. There is then a lag period before these cells can fully use the lactose and resume rapid growth. In both cases, the lag is due to the fact that β-galactosidase is not made if lactose is unavailable, or it is made only in small amounts if a preferred sugar, such as glucose, is also present. The cells have a mechanism for controlling the synthesis of β-galactosidase so that they only make it when it is useful for the cell (i.e., when lactose is available) and only in large amounts when it is the best available sugar. Those regulatory mechanisms for lactose use were initially discovered by François Jacob and Jacques Monod through a series of beautiful experiments involving a combination of genetics and biochemistry, work for which they were awarded the Nobel Prize in 1965. In the ensuing years, many more experiments by many different investigators have uncovered additional details, making the *lac* operon one of the best understood of regulatory circuits.

Figure 1-1 outlines the primary events of the regulatory system for lactose use. Four genes are indicated in the figure: *lacI*, *lacZ*, *lacY*, and *lacA*. *lacI* encodes the Lac repressor protein, which controls the expression of the other genes and is made at a low and nearly constant level. *lacZ* encodes the β-galactosidase required for using lactose. *lacY* is a β-galactoside permease gene that facilitates transport of lactose into the cell, and *lacA* encodes β-galactoside transacetylase. Those three genes are transcribed under the control of a single promoter into a common mRNA (this unit of genomic DNA is called an operon), from which they are translated into proteins independently. Between the *lacI* and *lacZ* genes are three binding sites labeled C, P, and O in Figure 1-1; these binding sites control the expression of the operon. In the figure, P is the promoter element bound by the RNA polymerase (yellow shape in Fig. 1-1) that transcribes the operon into mRNA. O is the operator at which the Lac repressor protein (red shape) binds when no lactose is available. Binding of the repressor to the operator excludes binding of the RNA polymerase to the promoter,

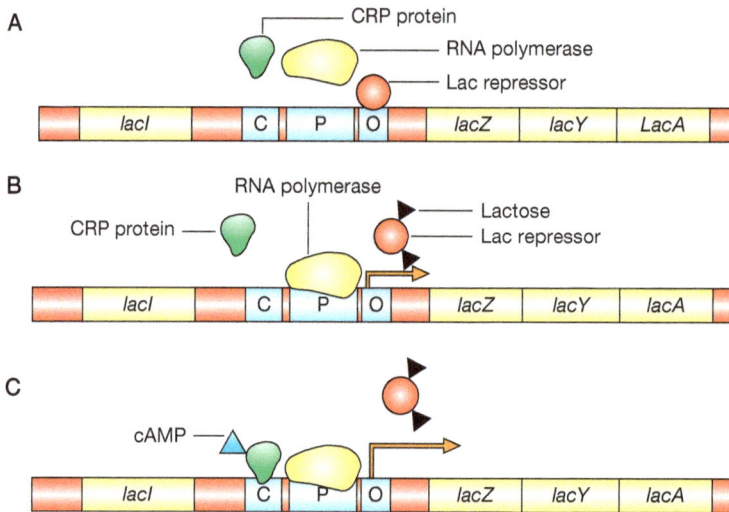

Figure 1-1. Diagram of the *lac* operator and surrounding genes. *LacI* is the gene for the Lac repressor, and the *lacZ*, *lacY*, and *lacA* genes (*lacZYA*), which are transcribed onto a single mRNA, are needed for use of lactose. The intergenic space between the *lacI* and *lacZ* genes controls the expression of the *lacZYA* mRNA through the occupancy of sites labeled C, P, and O, binding sites for the CRP protein, RNA polymerase, and Lac repressor, respectively. (*A*) When lactose is not available, the Lac repressor (red shape) binds to the O site and prevents transcription of *lacZYA* mRNA. (*B*) When lactose (black triangles) is available, it binds to the Lac repressor, preventing it from binding to the O site. Now, the RNA polymerase (yellow) can bind to the promoter P and lead to a low level of transcription (orange arrow). (*C*) When cAMP (blue triangle) levels are high, cAMP binds to the CRP protein (green), which then binds to the C site. CRP facilitates RNA polymerase binding to P and this leads to higher levels of *lacZYA* mRNA synthesis.

in which case no mRNA is produced (as in Fig 1.1A). When lactose is present, it binds to the repressor (Fig 1-1B; black triangles indicate bound lactose), which prevents it from binding to the operator. This allows the RNA polymerase to bind to the promoter and transcribe a small amount of the operon mRNA (orange arrow, Fig 1.1B, represents transcription initiation). C is the binding site for the CRP (green shape) protein (cAMP receptor protein, also known as CAP [catabolite activator protein]), which does not bind to DNA unless it is also bound by cyclic AMP (cAMP; blue triangle, Fig 1.1C). When cells are grown in glucose, cAMP levels are low, CRP does not bind to C, and the level of *lac* operon expression is low (Fig. 1-1B). When glucose is unavailable, cAMP levels rise, leading to CRP binding to C, which makes the promoter more active (orange arrow, Fig. 1-1C).

The *lac* regulatory system, composed of the three binding sites C, P, and O (all within a region of ~70 base pairs of DNA), and the proteins that bind to them (actually protein complexes, because CRP binds as a dimer, Lac repressor as a tetramer, and RNA polymerase is a protein complex with several subunits) illustrate many fundamental principles of gene regulation. Every gene requires a promoter, the entry point for RNA polymerase to transcribe it into mRNA, but the activities of promoters can have different intrinsic activities and can also be modulated by the activities of other proteins. The *lac* P promoter is relatively weak on its own, generating only low levels of *lac* operon expression. Binding at the adjacent C site by CRP can activate the promoter, leading to an ~40-fold increase in expression. Binding at the overlapping O site by the Lac repressor can essentially turn it off completely. Both of these regulatory proteins, Lac repressor and CRP, act as sensors of the environment. When lactose is unavailable, and it would be a wasteful expenditure of energy and material to synthesize the mRNA and proteins of the *lac* operon, the Lac repressor prevents their synthesis. But when lactose is available and binds to the Lac repressor, repression is eliminated and synthesis of the *lac* operon enzymes commences so that the available lactose can be consumed for cell growth.

The Lac repressor and CRP are typical of many bacterial TFs. They belong to the helix-turn-helix protein family in which an α-helical segment of these proteins interacts directly with the DNA sequences of their binding sites (see Chapter 3). They both dimerize (actually, the Lac repressor is a tetramer, or dimer of dimers) and bind to DNA sequences that are approximately palindromes (see Chapter 2), and their DNA-binding activities are modulated by small molecules, termed effector molecules (this is lactose in the case of the Lac repressor and cAMP in the case of the CRP). Note that the response to the effector molecule is opposite for these two proteins: the Lac repressor only binds the operator DNA when lactose is absent, whereas the CRP only binds to its binding site when cAMP is present. And this is not simply a distinction that the Lac repressor represses transcription and CRP activates transcription. Other repressors bind DNA only when the effector molecule is present, and other activators bind DNA when the effector molecule is absent. Thus, all combinations are possible: repressors and activators whose DNA-binding activity is either on or off in the presence of the effector molecule.

THE BACTERIOPHAGE λ GENETIC SWITCH

Another example of gene regulation that was discovered in the early days of molecular biology is the lysis/lysogeny decision of bacterial phage λ. Phage DNA can either merge into the host genome (lysogeny) and be multiplied in conjunction with the cells, or it can kill the cells (lysis) and in the process make many copies of itself ready to infect new cells. The choice primarily depends on two proteins competing for binding to three adjacent sites in the λ genome (Fig. 1-2A). These two proteins are the λ repressor and Cro. The genes for the two proteins are adjacent, on the λ chromosome, but are transcribed in opposite directions from a common intergenic regulatory region. That regulatory region contains five important elements: two promoters (P_R and P_{RM}) and three operators (OR1, OR2, and OR3). P_R represents the promoter

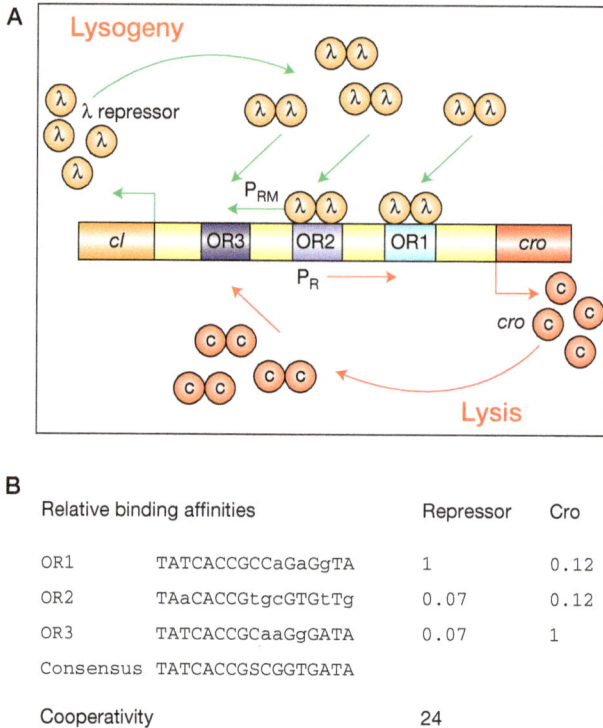

Figure I-2. (A) Switch region of the λ genome. The *cl* and *cro* genes, which encode the λ repression and Cro proteins, respectively, are separated by the intergenic region containing the binding sites OR1, OR2, and OR3 and the promoters P_R and P_{RM}, which drive expression of the *cro* and *cl* genes, respectively. Both proteins dimerize and can bind to all three operators. Binding by the λ repressor at OR1 and OR2 blocks transcription from P_R and activates transcription from P_{RM}. Binding at OR3 by either protein blocks transcription from P_{RM}. (*A*, Reproduced, with permission, from Macmillan Publishers Ltd.) (*B*) DNA sequences at the three binding sites and their palindromic consensus sequence (S means "C or G"). The relative binding affinity (ratio to maximum) of each protein for each binding site is shown and the cooperativity for λ repressor binding to adjacent sites, where the total affinity is the product of the individual affinities and the cooperativity factor.

for expressing the Cro protein; it overlaps OR1 and OR2, so that when they are occupied by either protein, P_R is turned off and Cro is not expressed. P_{RM} is the promoter for expressing λ repressor (RM is for repressor maintenance); another promoter for the λ repressor, P_{RE} (RE is for repressor establishment [off to the right are not shown in Fig. 1-2]) is responsible for the initial expression of repressor. OR3 overlaps P_{RM} so that when it is occupied by either protein, P_{RM} is off and λ repressor is not made. OR1 and OR2 do not overlap P_{RM} and when they are occupied by the λ repressor, they activate P_{RM}, leading to increased synthesis of the repressor. Thus, in fact, the λ repressor can act as either an activator or a repressor of transcription, depending on where it binds relative to the promoter sequence.

OR1, OR2, and OR3 are all binding sites for both Cro and λ repressor, and these two proteins compete for occupancy of those sites. However, the three operators have similar, but distinct, DNA sequences (Fig. 1-2B), and those differences lead to opposite preferences for binding. λ repressor prefers binding to OR1 and, through cooperativity (see Chapter 5), also binds OR2 with high affinity. When the λ repressor is initially synthesized from the P_{RE} promoter, it first binds to OR1 and OR2 cooperatively, repressing P_R and activating P_{RM}, which leads to increased expression of the λ repressor. As its concentration increases, it then binds to OR3, which turns off P_{RM}. This is a very compact feedback loop that helps maintain λ repressor at a nearly constant concentration in the cell. When it is present in low amounts, it activates its own synthesis until it reaches the appropriate amount, at which point it turns off its own synthesis. Cro binds preferentially to OR3, so it initially functions to turn off P_{RM} and the expression of the λ repressor. At high concentrations, Cro will also bind to OR1 and OR2 and repress its own synthesis, but its primary role is in counteracting the effect of the λ repressor.

Cro and the λ repressor are each initially expressed at low levels, but their competition for binding to the intergenic region containing operators OR1, OR2, and OR3 leads to two possible outcomes: one of them wins the competition and continues to be expressed, and the other is turned off. If they merely controlled each other's expression, there would be no point to the competition; in fact, however, the winner of the competition dictates the expression of the rest of the λ genome. This regulatory circuit (the two proteins and their binding sites) functions as a bistable genetic switch; only one protein can win the competition and turn off the expression of the other protein, and the winner controls whether the phage becomes lysogenic or lytic.

The λ genome is more than 48 kb pairs in length and codes for at least 60 genes. Most of those genes are required for the lytic state in which they serve to take over the cell, make many new copies of the phage genome, and package it into many new phage particles. Eventually, the cell bursts, or lyses, and releases the new phage that can now infect new cells. The lysogenic state requires many fewer genes. It initially requires the expression of Int, a protein that functions to integrate the λ genome into the host, and the λ repressor to keep all of the other genes turned off. Once integrated and in the lysogenic state, only the repressor expression is need to maintain that state. In fact, if the host cell becomes stressed in various ways, it initiates an SOS response characterized by activation of the RecA protein. The phage takes advantage of this system because RecA will lead to cleavage and inactivation of the λ repressor. That stimulates the dormant phage to escape from the host by expressing the Xis protein, which excises the λ genome from the host chromosome and expression of Cro protein leads to expression of all the other lytic genes, replication of the phage genome, and release of many new phage particles. Thus, although the choice of the lytic pathway is irreversible, the lysogenic pathway can be reversed if the host cell is under stress and may not survive.

The description above is a simplification that omits some of the steps involved in the choice between lysis and lysogeny, but it does illustrate some principles that were not evident from the *lac* operon example. One principle is autoregulation of the λ repressor. The fact that the λ repressor both activates and represses its own synthesis means that its level can be tightly controlled in a simple and elegant feedback mechanism. The expression of many proteins, not just TFs, is autoregulated, and any step in the synthesis process may be subjected to regulation. In most cases, the regulation is only through repression: Proteins turn themselves off when sufficient quantities exist, and the activation is provided by another mechanism.

Another principle evident from the λ example is that competition among regulators can lead to a stable (although possibly reversible) switch. In the case of the *lac* operon, there is competition between RNA polymerase and the Lac repressor, but in the case of λ, the competition is directly between two TFs, only one of which can win the competition and the winner decides the fate of the infection (whether it becomes lysogenic or lytic). Although this genetic switch is particularly simple, it illustrates how alternative fates of cells can be directed by the interactions of TFs with DNA.

The third principle involves the fact that both Cro and the λ repressor bind to the same three operators with slightly different sequences. This highlights an important fact returned to throughout this volume—that TFs do not bind to just one sequence but rather to sets of similar sequences, and their binding affinities vary depending on the exact sequence. This results in quantitative differences in how well TFs bind to different sites and allows for the switch mechanism found in λ phage and many other regulatory systems.

REGULATORY ACTION AT A DISTANCE AND COMBINATORIAL LOGIC

The same principles about sequence-specific TFs regulating gene expression hold in eukaryotic organisms, but there are also a number of important differences. One is that the DNA of eukaryotes is confined to the nucleus and protein synthesis occurs in the cytoplasm, whereas bacteria have no nucleus and both RNA and protein synthesis (transcription and translation) occur together; in fact, a nascent mRNA can be translated into protein as it is being synthesized. In eukaryotes, TFs have to transit into the nucleus to effect transcription, and one form of regulation is to sequester the factor in the cytoplasm until a signal is received that targets it to the nucleus. Many signaling cascades are initiated by a receptor on the surface of the eukaryotic cell; the receptor binds to a specific compound, which then releases, directly or indirectly, a TF to enter the nucleus to effect transcription. In addition, the RNA polymerase itself is composed of a much larger complex, with many more associated proteins, in eukaryotes. Another important difference is that eukaryotic genomes are larger (generally, much larger) than bacteria. The yeast *Saccharomyces cerevisiae* has a genome of ~12 million base pairs (12 Mbp, or megabase pairs), only slightly more than twice as large as the bacterium *E. coli*. But most eukaryotes, especially the multicellular ones with which we are most familiar, have much larger genomes. Commonly studied model organisms such as worms and flies have ~100 Mbp of genomes, and most mammals, including humans, have genomes of ~3 Gb (billion or gigabase pairs). Although eukaryotic genome size can be nearly 1000 times larger than bacteria, the number of genes is usually only four or five times larger. This results in a much larger fraction of the genome not coding for proteins. Some of that noncoding DNA is devoted to regulatory regions that control the expression of genes, some is expressed RNA that has other roles in the cell besides coding for protein (so-called noncoding RNA [ncRNA]), and some is commonly referred to as "junk" DNA, which mostly means that if it has a function, we do not know what it is yet. Eukaryotic genes also contain large segments known as introns, parts of pre-mRNA transcripts that are removed by splicing to generate the final mRNA sequence. It is common for pre-mRNAs to have multiple alternatively spliced forms leading to multiple different proteins for a single gene. So although the number of genes in multicellular eukaryotes may be only four or five times as many as that in simple bacteria, the number of proteins encoded may be much greater.

One of the most important differences is the accessibility of genomic DNA in prokaryotes versus eukaryotes. Although there are many proteins in bacteria that bind to DNA, and many of them compete for binding to the same regions, we generally believe that essentially all of the DNA is accessible to TFs. In eukaryotes, most of the DNA is wrapped around histone proteins to form units called nucleosomes, which are the chromatin of the nucleus (Fig. 1-3). Nucleosome-free regions are accessible to TFs and a few proteins can bind to DNA that is wrapped around a nucleosome. In addition, all of the chromatin is dynamic, and thus some regions are accessible at certain times and in certain cells but inaccessible at other times or in other cells. The regulatory regions that are active in any specific cell are those bound by TFs in that cell; they can be located far away on the genome from the genes that they regulate.

Although regulation at a distance has been found in bacteria, generally the regulatory regions in bacteria overlap or are adjacent to the promoters, as described for both the *lac* operon and the λ switch. In eukaryotes, action at a distance is the general rule. There are factors that bind near to the promoter, where RNA polymerase binds and transcription starts, but regulatory regions can occur far upstream of or downstream from (relative to the direction of transcription) the gene. These regions, called enhancers, often function to activate transcription (Fig. 1-3), but other regulatory regions can also repress transcription, perhaps by blocking the effect of an enhancer.

Another common feature of eukaryotic gene regulation is that TFs typically work in combinations, usually with several TFs coordinately controlling gene expression (Fig. 1-3). This can occur in bacteria, but it is much more common for a bacterial gene to be controlled by a single factor or perhaps by the competition between pairs of factors. In eukaryotes, the general case is combinations of TFs binding to enhancers

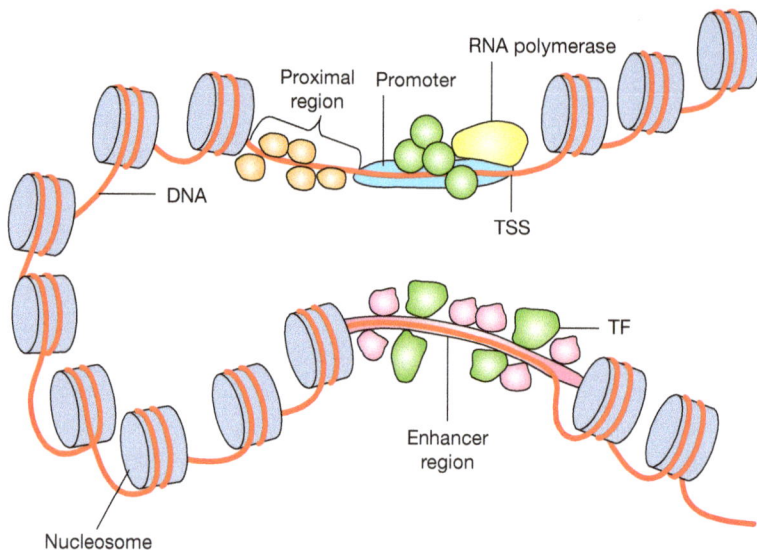

Figure 1-3. Example of a eukaryotic gene and its control regions. Most of the DNA is wrapped around nucleosomes, but there are nucleosome-free regions where regulatory proteins can bind. Shown are the promoter region, where the RNA polymerase complex binds and initiates transcription, and a promoter-proximal region that includes binding sites for some factors that facilitate transcription. Also shown is an enhancer region, which binds other factors that activate transcription. The enhancer region is some distance away on the chromosome but is brought into close proximity by looping of the intervening DNA. (TSS) Transcription start site.

in a coordinated fashion. There may be several distinct enhancers for a specific gene, each controlling its expression in certain cells or in certain conditions, but in every case the fundamental aspects of proteins recognizing and binding to specific DNA sequences follow the same principles. These principles are the primary focus of this volume.

OUTLINE OF THE BOOK

The study of protein–DNA interactions has proceeded using three distinct approaches: analysis of the structure of DNA and protein separately and in complexes; biochemical and biophysical analyses of complex formation including both the equilibrium and kinetic properties; and bioinformatics approaches, based largely on statistical analyses of large collections of interaction data. Structural analyses include not only the shape and geometry of the interacting partners but also the chemical nature of the interface— the noncovalent bonds that provide the affinity and specificity of the interaction. Biochemical and biophysical approaches determine affinity and specificity quantitatively and can separate binding energy contributions into various categories that provide insight into the molecular mechanisms—between entropic and enthalpic contributions, electrostatic and polar and nonpolar contributions, and sequence-specific and nonspecific interactions. Bioinformatic analyses determine statistical properties of classes of interactions that facilitate inferences about molecular mechanisms and predictions of novel interactions. Each approach brings a unique perspective to the problem. Most researchers of protein–DNA interactions are experts in one of these areas with a less comprehensive knowledge of the others. In this introductory volume, those experts are not likely to find new information or insights in the area of their own specialty, but they may benefit from the descriptions of the other areas. Readers not expert in any area will gain a wide perspective on what is known about protein–DNA interactions, how such knowledge is obtained, and how it is useful for a broader understanding of biology in general.

Technological advances during the last 60 years have fueled the enormous expansion in our knowledge of all biological processes. Every discipline relies heavily on computational analyses, and the exponential increase in computing power, as forecast in Moore's Law, has been essential for much of this progress. Advances in other methods of instrumentation have also been enormous, allowing for increased throughput, more sensitive data collection, and in many cases the collection of new types of data that was not previously possible. This is nowhere more important than in techniques that allow for the sequencing of DNA, only invented in the late 1970s, and for the synthesis of designed DNA sequences, invented a few years later. Both have undergone tremendous improvement in succeeding years, resulting in much greater throughput and efficiency, decreased error rates, and ever-decreasing costs. Those methods have enabled a wide variety of experiments to be undertaken that were inconceivable not long ago.

Protein–DNA interactions are integral to many areas of biology: cellular responses to the environment, differentiation and development, normal phenotypic variations due to altered gene expression and misregulation associated with many diseases, and many areas of biotechnology and synthetic biology. Our knowledge of protein–DNA interactions has increased enormously in recent years but there is still much more to be learned. The primary goal of this volume is to provide an introduction to this topic from multiple perspectives that summarizes much of what is known and facilitates an integrated understanding of protein–DNA interactions.

FURTHER READING

Judson HF. 1995. *The eighth day of creation: Makers of the revolution in biology* Penguin, London Expanded edition: 1996, 2005. Cold Spring Harbor Laboratory Press, Cold Spring Harbor, New York.
   Illuminates history of the early decades of molecular biology, including fascinating glimpses into the personalities of the pioneering scientists.

Ptashne M. 2004. Phage λ revisited. In *A genetic switch*, 3rd ed. Cold Spring Harbor Laboratory Press, Cold Spring Harbor, New York.

Excellent overview of the phage λ genetic switch, including many more details not present in this chapter. The previous two editions are also excellent.

Ptashne M, Gann A. 2002. *Genes & signals* Cold Spring Harbor Laboratory Press, Cold Spring Harbor, New York.

Another excellent overview of gene regulation, including the *lac* operon, the phage λ switch, and several examples from eukaryotic systems.

CHAPTER 2

The Structure of DNA

THERE HAVE BEEN THREE GREAT STRUCTURAL DISCOVERIES IN the history of science: (1) the sun-centered structure of the solar system by Copernicus, (2) the multiple electron orbits of the atom by Bohr, and (3) the double-helical base-paired structure of DNA by James D. Watson and Francis H.C. Crick. In each case, the structure was simple and elegant and provided an obvious explanation for previously mysterious observations: the path of the planets through the sky, the discrete emission spectra of atoms, and the replication of the genetic material during cell division. In their first paper on DNA structure, Watson and Crick (1953a) observed that "The sequence of bases on a single chain does not appear to be restricted in any way. However, if only specific pairs of bases can be formed, it follows that if the sequence of bases on one chain is given, then the sequence on the other chain is automatically determined." And then, in the now famous and perhaps most understated sentence in the scientific literature, they continued, "It has not escaped our notice that the specific pairing we have postulated immediately suggests a possible copying mechanism for the genetic material." The details of that model had to await their following paper, 1 month later (Watson and Crick 1953b).

BASE PAIRS AND THE DOUBLE HELIX

The primary features of DNA structure are diagrammed in Figure 2-1A. The bases on each of the two strands of DNA are covalently linked through a sugar–phosphate backbone, and the two strands are held together by noncovalent hydrogen bonding between the nucleotide bases on each strand. The pairing rules that adenine (A) pairs with thymine (T), and guanine (G) pairs with cytosine (C), dictate that the two strands are complementary—the sequence of bases on one strand specifies precisely the sequence on the other strand. When separated, each strand provides a template on which the other strand can be synthesized, allowing faithful duplication of the DNA double helix that can be passed on to the daughter cells (Fig. 2-1B). Strand

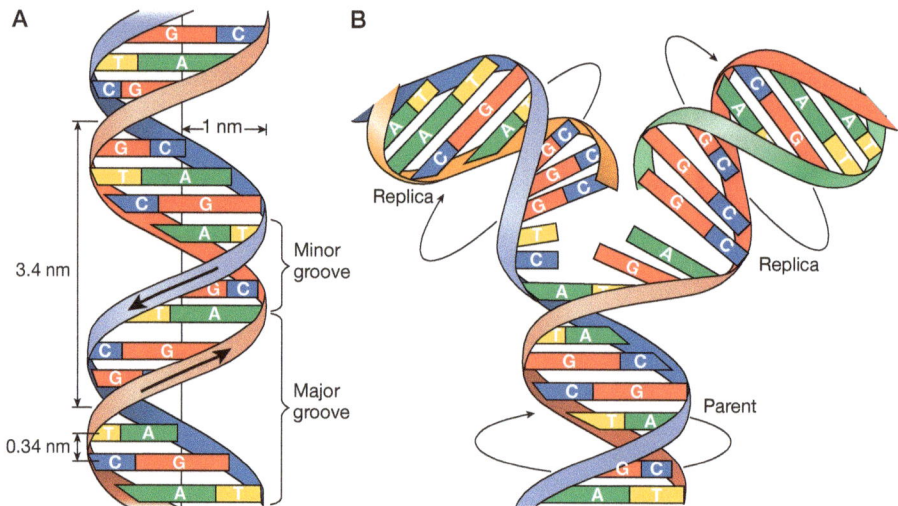

Figure 2-1. Structure of DNA. (*A*) Diagram showing base pairs with 0.34-nm vertical spacing and 10 base pairs per turn of the helix for a total pitch of 3.4 nm. Strands are antiparallel and have a major, or wide, groove and a minor, or narrow, groove where the edges of the base pairs are accessible. The base pairs are perpendicular to the helical axis, and the width of the helix is ~2 nm. (*B*) The replication process is evident from complementarity of the two strands. When the two strands are separated, as shown at the top of the figure, a new strand can be synthesized using the old one as a template. The result is two dsDNAs with identical sequences.

separation also occurs during transcription so that one strand of the DNA can be copied into RNA (see Box 2-1). Each strand has a polarity, defined by the sugar–phosphate backbone, and the two strands are antiparallel to each other. The two "grooves" separating the backbones are of different widths, the wider one referred to as the "major groove" and the narrower one as the "minor groove." (See below for more details on DNA grooves.) The height of each base pair is approximately one-third of a nanometer and the rotation between adjacent pairs is ~36°, resulting in a right-handed helix with a pitch of 10 bp in 3.4 nm. The entire width of the helix is ~2 nm. The length of a complete DNA molecule varies considerably among different species. The bacterium *Escherichia coli* has ~5×10^6 base pairs in its chromosome, for a total length of ~1.7 mm, compared to the length of an *E. coli* cell of ~2 μm (see Box 2-2). The human genome contains ~3×10^9 base pairs, divided into 23 chromosomes (for the haploid genome, the diploid cell has 46 chromosomes, 22 pairs of autosomes, and either two X chromosomes for females or an X and Y chromosome for males). The average length of a chromosome is ~43 mm if stretched out linearly compared to a cell nucleus that is typically a few micrometers in diameter.

In addition to providing an explanation for genetic replication, the structure of DNA also has implications for the structure of a gene. It was known since the experiments by Oswald Avery et al. (1944) that DNA was the genetic material and that there

Box 2-1. DNA Melting

DNA melting, or denaturation, refers to the separation of the two strands of double-stranded DNA (dsDNA). In a cell, the strands are separated during replication so that the new DNA strands can be synthesized (see Fig. 2-1B). The strands are also separated during transcription when RNA polymerase opens up a "transcription bubble" so that it can synthesize an RNA strand complementary to one of the DNA strands (Box Fig. 2-1). The bubble closes soon after the RNA polymerase moves on, so the melted region is of limited size, as is the case of DNA replication. There are also short stretches of single-stranded DNA (ssDNA) generated during the repair and recombination processes. All of these melted regions are generated by proteins, are very limited in size, and do not last long, so that the vast majority of DNA inside a cell is in its normal dsDNA state.

In vitro (in a test tube), the dsDNA can be separated completely into two single strands. This is accomplished most easily if the DNA is fragmented into pieces that are much smaller than whole chromosomes; DNA can be readily fragmented by vortexing, sending it through a small needle at high velocity, sonication, or other methods. Heating dsDNA is the easiest way to denature it. Depending on the length and composition (GC-rich DNA is more stable than AT-rich DNA), the melting temperature is typically between 80°C and 90°C in normal buffers. In extremely low-salt concentrations, DNA will melt at lower temperatures because it is stabilized by positive ions such as Na^+ and Mg^{++}. Other chemicals can also facilitate denaturation.

THE VALUE OF MELTING DNA

In the late 1960s, Roy Britten and colleagues discovered repetitive DNA through experiments that denatured and renatured DNA (Britten and Kohne 1968). Sheared DNA that is denatured by high temperature will renature (reforming the dsDNA) when the solution is cooled. How fast it renatures depends on the concentration, because each single strand has to find its partner to reform the dsDNA, or at least a sequence that is very similar so that it can form a

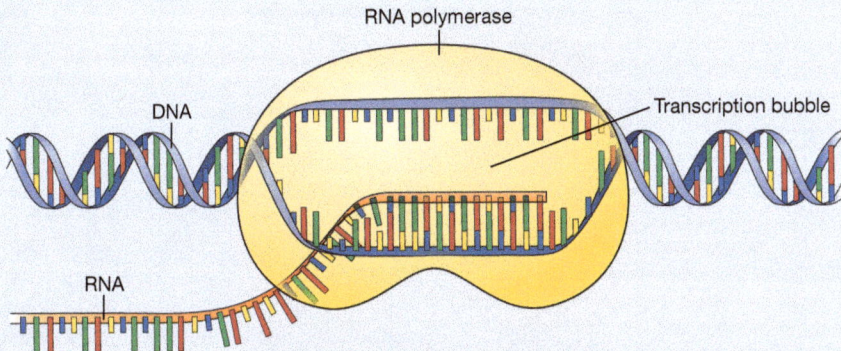

Box Figure 2-1. Transcription bubble showing RNA polymerase in the melted open region of DNA with new complementary RNA strand emerging.

Box 2-1. (*Continued.*)

nearly perfect dsDNA hybrid. For a given total amount of DNA, the renaturation rate then serves as a means to determine the complexity of the DNA—the number of different sequences. For simple genomes such as bacteria and viruses, there is a linear relationship between the time required for renaturation and the length of the genome. For example, the phage T4 has a genome that is ~30 times smaller than that of *E. coli*, and if the same total amount of DNA is denatured the T4 genome will renature ~30 times faster than the *E. coli* genome. When the same experiment was tried on larger genomes, including invertebrates, plants, mammals, and many others, a surprise finding was that a portion of those genomes renatured very quickly, suggesting that it was present in hundreds to thousands of copies per genome. In the intervening years, we learned by a variety of techniques, and now by direct DNA sequencing of the genomes, that ~80% of the human genome contains many families of repetitive elements, some occurring hundreds and others thousands of times in the genome, just as suggested from the original renaturation experiments.

Denaturation of dsDNA is also part of a very important experimental procedure called polymerase chain reaction (PCR) (Box Fig. 2-2). This invention by Kary Mullis (1990) won

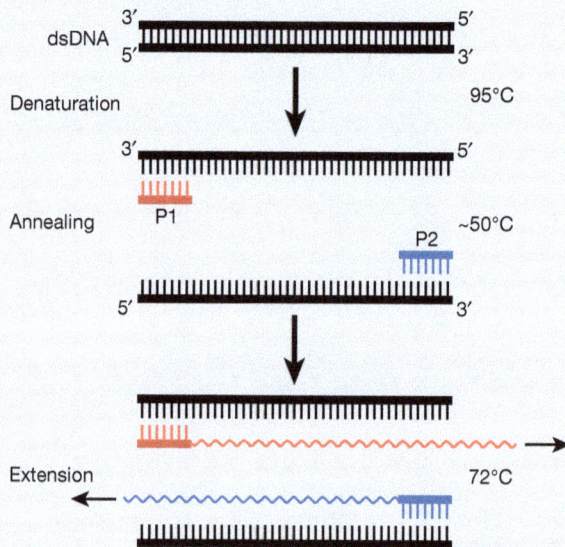

Box Figure 2-2. Polymerase chain reaction. dsDNA is heat denatured (Denaturation). The reaction is cooled and primers complementary to the end of each of the two original target DNA strands are annealed (Annealing). Primers (P1,P2) are extended (Extension) by a DNA polymerase to create two copies of the original dsDNA. The entire process is repeated as many times as needed (or until the primers are exhausted), with each round doubling the amount of DNA. Millions of copies of the target region of DNA are produced. Note that in subsequent cycles of PCR, the DNA generated is also used as a replication template, so that subsequent cycles result in an exponential amplification of the DNA template. (Modified from Florida Museum of Natural History; http://www.flmnh.ufl.edu/cowries/amplify.html.)

him the Nobel Prize in Chemistry (shared with Michael Smith) in 1993. PCR provides a means of exponentially amplifying a specific piece of DNA (the segment between the primers), making it possible essentially to produce unlimited amounts of particular DNA sequences. And by positioning adapters, which contain the binding sites for the primers, on the ends of all DNA sequences in a mixture, PCR allows large-scale amplification of the entire mixture. The same result can be accomplished using random primers that will cover the entire genome. PCR, and various adaptations, is critical for many different processes now common in biological research and biotechnology, including the sequencing of genomes from single cells.

Box 2-2. Scaling *E. coli* up to Human Dimensions

It is sometimes useful to imagine shrinking oneself in size so that the inside of a cell can be examined in detail, or equivalently, expanding the size of the cell up to human dimensions. A convenient scale factor is 10^7, which converts a nanometer, the scale relevant for DNA, proteins, and many cellular structures, to a centimeter, the size of many common objects. Consider an *E. coli* bacterium, which is typically ~2 μm long and ~0.5 μm in diameter. That would scale to a room with dimensions of ~20 m by 5 m, comparable to many classrooms or auditoriums.

Let us consider that a classroom represents the inside of an *E. coli* cell. At this scale, the width of DNA is ~2 cm, approximately twice the width of a typical electrical cord. The length of the *E. coli* genome would be ~17 km and must be completely contained within our classroom, which requires that it be wrapped around the room many hundreds of times. When grown in ideal conditions, *E. coli* can double approximately every 30 min, and each doubling requires a complete replication of the DNA. There is a single origin of replication for *E. coli* DNA, and DNA synthesis moves in both directions around the circular chromosome until the two replication forks meet at the terminus. In our cell, each replication fork moves at ~17 km/h, and as it travels along the DNA it separates the two strands and makes a highly faithful copy of each, so that at the end there are two complete genomes within the cell (which itself has about doubled in length, with a nearly constant diameter, during the same period).

Inside the cell are also ~10,000 ribosomes, each the approximate size of a volleyball. There are a similar number of mRNA molecules, copies of short stretches of the DNA, averaging ~10 m in length. There are several hundred RNA polymerase complexes busy copying DNA into segments of mRNA, moving at a relatively leisurely pace (compared to the DNA polymerase) of ~1 km/h, with ribosomes synthesizing proteins at a similar rate. In addition are ~1 million protein molecules, typically 3–6 cm in diameter (marble to golf ball sized), with many of them forming complexes that measure several centimeters in diameter. At the same time, the cell is monitoring the outside environment, adjusting the synthesis rate of different proteins to adapt to that environment and steer the entire cell (yes, it can move; maybe we should think of it as a giant bus instead of a classroom) in directions that it anticipates to represent a better environment. Communication with the outside is through very small portholes, not the doors and windows of our classroom. They can selectively

---

Box 2-2. (*Continued.*)

---

import and export individual sugar molecules, amino acids, and other small molecules such as water and various ions, which are mostly <1 cm in diameter.

Overall, the inside of the cell is a very crowded and busy place. When operating at its maximum rate, it can duplicate all of the components and generate an entirely new cell complete with its own genome within ~30 min. This is done with incredible accuracy; typically, less than one mutation is introduced into the new genome at each generation. The TFs that are the main topic of this book have a critical role in cellular processes by controlling which genes are expressed and in what amount, under the various conditions that the cell encounters. Understanding how they distinguish among different DNA sequences is fundamental to modeling cellular behavior.

---

was a correspondence between genes and proteins; the "one gene–one enzyme" hypothesis had been proposed by George Beadle and Edward Tatum (1941) and would result in a Nobel Prize in 1958. But how DNA could specify a protein was unknown. There were many hypotheses, such as the DNA forming a template or scaffold on which the amino acids would condense into the protein structure. But the structure as elucidated by Watson and Crick immediately suggested instead that DNA is a digital information carrier, using a four-letter alphabet, with the instructions for a gene encoded in the sequence. This further implied that there had to be a decoding process that translated the DNA sequence into the protein sequence, which has a 20-letter alphabet (see Chapter 3). Within the next decade, the decoding mechanism, including both transcription of DNA into RNA and translation of RNA into protein and the "machines" that carried them out (RNA polymerase and the ribosome), had been discovered and the full extent of the genetic code had been deciphered. The notion of molecular paleontology, through the comparison of protein and DNA sequences between and within species, also emerged during the next decade and gave clear insights into the process of evolution at the molecular level.

PROTEIN–DNA INTERACTIONS AND GENE REGULATION

By the early 1960s, the basic principles of gene regulation had been established (see Chapter 1). Some proteins function as *trans*-acting factors that bind to specific *cis*-acting regulatory sites to control the transcription of the adjacent genes. The pioneering studies of lactose use in *E. coli* and the lysis/lysogen switch in phage λ established the paradigm, but it was soon discovered that such systems are ubiquitous and not just in bacteria. Although the specific details can be more complicated and there are often additional layers of regulation, the basic principles have remained largely intact. Proteins that function as transcription factors (TFs) bind to specific regions of DNA and affect the expression of genes that are controlled by those regulatory sites (see

Chapter 1). In bacteria, the regulatory sites are typically quite close to, and generally upstream of or overlapping, the start of transcription. They can be quite far away in eukaryotic cells, functioning as "enhancers" that can up-regulate gene expression or "repressors" that can down-regulate gene expression, depending on the conditions and factors that bind. Complexity arises as a result of combinatorial interactions among factors, the competition for binding to DNA with chromatin structure, epigenetic modifications to the DNA and chromatin, looping of DNA, and perhaps other processes. But at a fundamental level, the regulatory machinery must distinguish among different segments of the genome so that the appropriate genes are turned on and off, and this requires that proteins decipher the information in the DNA through interactions that depend on the sequence.

ACCESSIBLE SURFACES OF BASE PAIRS

Figure 2-2 shows the structure of the backbone, which is the same for each type of base, and the chemical details of each type of base pair. The backbone is negatively charged on the phosphate groups and is neutralized in solution and in the cell by cations such as sodium or magnesium. In fact, at very low-salt concentrations, DNA is less stable and the two strands separate (the DNA "melts") at a lower temperature than it does at physiological salt concentrations. This negatively charged ladder allows for strong interactions with proteins, especially for positively charged amino acids such as arginine and lysine (see Chapter 3). Those interactions provide affinity for the protein–DNA interaction but not specificity unless the shape of the backbone is altered (see next section).

The major groove contains the most distinguishing features for base pairs. The positions of hydrogen-bond donors and acceptors can distinguish an A-T pair from a G-C pair and the orientation of the G-C base pair. For example, a C-G pair has donor (D) and acceptor (A) atoms in the order DAA across the major groove, whereas in the reverse orientation a G-C pair would have AAD. For both T-A and A-T, the order is ADA, but in addition there is the hydrophobic methyl group on the 5 position of T that can be used to distinguish the orientation through nonpolar (van der Waals) contacts rather than hydrogen bonds. The major groove can be visualized as a helical ladder with the rungs representing an array of chemical groups (Fig. 2-3) that are exposed to interact with the surface of protein molecules with similar types of chemical groups (see Chapter 3). When proteins come in close contact with the DNA such that their hydrogen-bond acceptor groups are opposite to the DNA's donor groups, or vice versa, the resulting hydrogen bonds between the DNA and protein increase their binding affinity. Unfavorable interactions result when two donor groups, or two acceptor groups, are in close contact. The exact amount of binding energy resulting from a specific protein–DNA hydrogen bond will depend on the distance and orientation between them, but in general, proteins will bind with higher affinity to DNA sequences with more favorable contacts than to those sequences with fewer

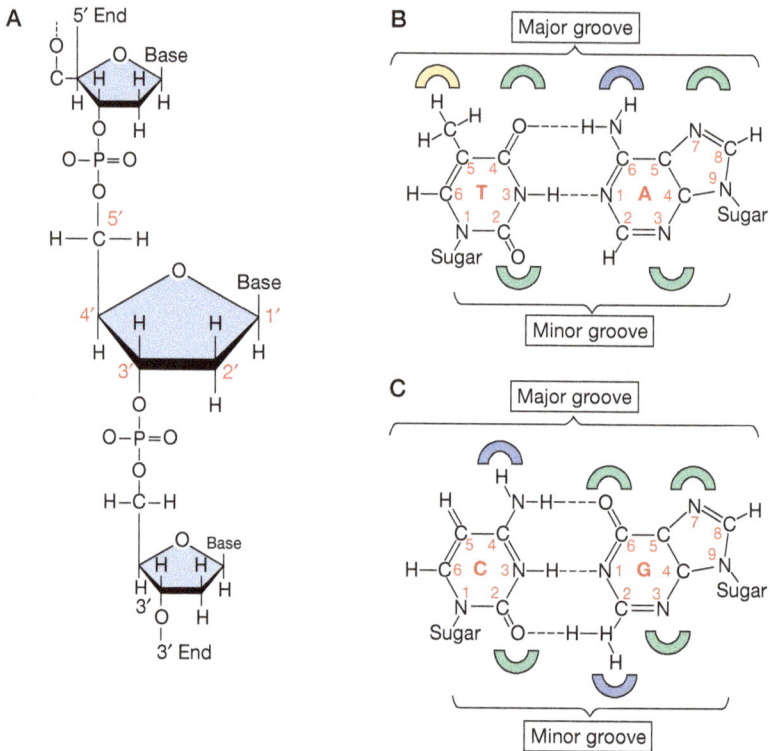

Figure 2-2. Chemical properties of DNA. (A) The backbone of DNA, which links the bases together, is composed of a chain of the sugar deoxyribose linked through phosphate groups. The chain has a polarity with one end referred to as 5′ and the other as 3′, based on the number of sugar carbon atoms. Each phosphate group is negatively charged. (B) Chemical structure of an A-T base pair with numbers assigned to the main atoms of each base. The link to the sugar is shown, and hydrogen bonding between the bases is indicated with dotted lines. Hydrogen-bond donors (blue) and acceptors (green) are shown in both grooves, as is the methyl group on T (yellow). (C) Chemical structure of a G-C base pair with numbers assigned to the main atoms of each base. The link to the sugar is shown, and hydrogen bonding between the bases is indicated with dotted lines. Hydrogen-bond donors (blue) and acceptors (green) are shown in both grooves.

favorable, and more unfavorable, contacts. van der Waals contacts also contribute to the binding affinity, and there may also be water atoms at the interface between the protein and DNA that form bridges of hydrogen bonds (see Chapter 4). The minor groove also has hydrogen-bond acceptors and, in the case of a G-C pair, a hydrogen-bond donor, but the orientation for each pair is nearly identical. The T-A pair and an A-T pair each have *AA* in the minor groove, and both a C-G pair and a G-C pair have *ADA*. In addition, the minor groove is narrower than the major groove, making it less accessible to the surface of a protein. Even so, the minor groove can also be visualized as a helical ladder with chemical groups arrayed on the rungs and can be used to help

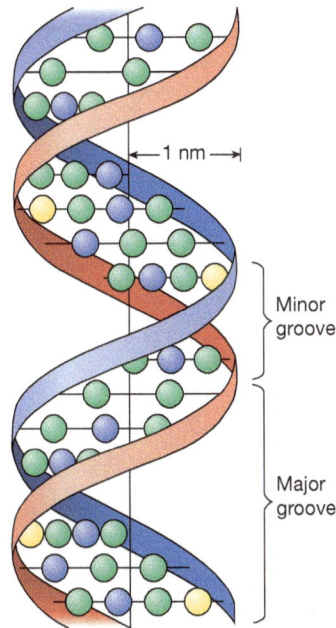

Figure 2-3. Structure of dsDNA showing chemical groups in both groups. This is the same image as that in Figure 2-1A but with base-pair labels replaced by hydrogen-bond donors (blue), acceptors (green), and methyl groups on T (yellow) as they appear in both the major and minor grooves of DNA.

distinguish different sequences through hydrogen-bond interactions, much as the major groove (Fig. 2-3). This use of chemical groups in the major and minor grooves is often referred to as "direct readout" because it uses the specific H-bonding and van der Waals signatures of each base pair to distinguish among different sequences.

Another important feature of DNA structure is the potential symmetry of DNA sequences. Because the backbones of the two strands are antiparallel, there is no structural distinction between the two strands; for example, one cannot be called the "top" and the other the "bottom" strand except in relation to some other reference mark. Commonly, one strand is referred to as "Watson" and the other as "Crick" just to provide a label for each, but the distinction is arbitrary and used only for convenience. This means that within DNA, there exist sequence segments that are symmetric, meaning that they are equivalent when read from either direction on dsDNA. These are commonly referred to as DNA palindromes, but unlike palindromic words, they are not read the same forward and backward on the same strand; rather, they are the same if each strand is read in the same direction, such as 5' to 3', which is the convention. For example, the sequence 5'-GATC-3', if read on the other, complementary strand in the same direction (from 5' to 3'), would be the same (5'-GATC-3'). Because many TFs form a complex with themselves, called homodimers, a single factor will often bind to symmetric sequences. This doubles the length of the binding site and the amount of information that can used to distinguish among different sites without increasing the size or complexity of the factor itself.

MODIFIED BASES

Although most bases in DNA are the standard A, C, G, and T, there are also some modified bases. In vertebrate organisms, the C of a CG dinucleotide (often referred to as CpG to indicate that they are on the same strand and not base paired, but here we indicate base pairs with a "−," so we stay with the convention of writing CG to indicate the sequence on one strand) is frequently methylated at the 5 position, the same position as the normal methyl group of T (Fig. 2-2B,C). Because CG is palindromic, the Cs on both strands are methylated. This occurs postreplication; the base inserted by DNA polymerase is a normal C, but it is enzymatically modified to become 5-methyl-C. In bacteria, a common modification is N6-methyl-A. This is the N connected to carbon 6 of A and involved in base pairing (Fig. 2-2B), with the methyl group replacing the H that is not involved in the hydrogen bond. Again, this modification is added postreplication and in fact is used by bacteria to distinguish the newly synthesized strand from the parental strand to aid in the correction of synthesis errors. These modifications can affect the affinity of proteins that bind to the DNA sequences. Some proteins will be indifferent and bind with the same affinity regardless of whether a base is modified, but other proteins may be highly sensitive to modified bases. In mammals, the 5-methyl-C is an "epigenetic" mark that influences the interactions of the DNA with proteins, including the overall chromatin structure. Such epigenetic changes are not permanent changes to the DNA sequence; they can be reversed and can serve as an additional level of regulation beyond recognition of the sequence alone. A few other base modifications are observed only rarely or in the genomes of some specific species. For example, the bacteriophage T4 incorporates 5-hydroxymethyl-C into its DNA during replication and is then enzymatically glycosylated at nearly every C in the genome. Such modifications can have large effects on the affinity of some DNA-binding proteins.

SEQUENCE-DEPENDENT VARIATION IN THE DOUBLE HELIX

The double-helical structure shown in Figure 2-1A is an average, or canonical, structure; however, individual segments may differ in various ways, and the differences are dependent on the sequence. Each base pair can move slightly or rotate slightly, relative to the helical axis and with respect to one another (Fig. 2-4A) and also with respect to the neighboring base pairs (Fig. 2-4B) (Olson et al. 2001). The energetic cost of those movements depends on the sequence, so that different sequences will have higher propensities than others to adopt specific structural deformations. Besides

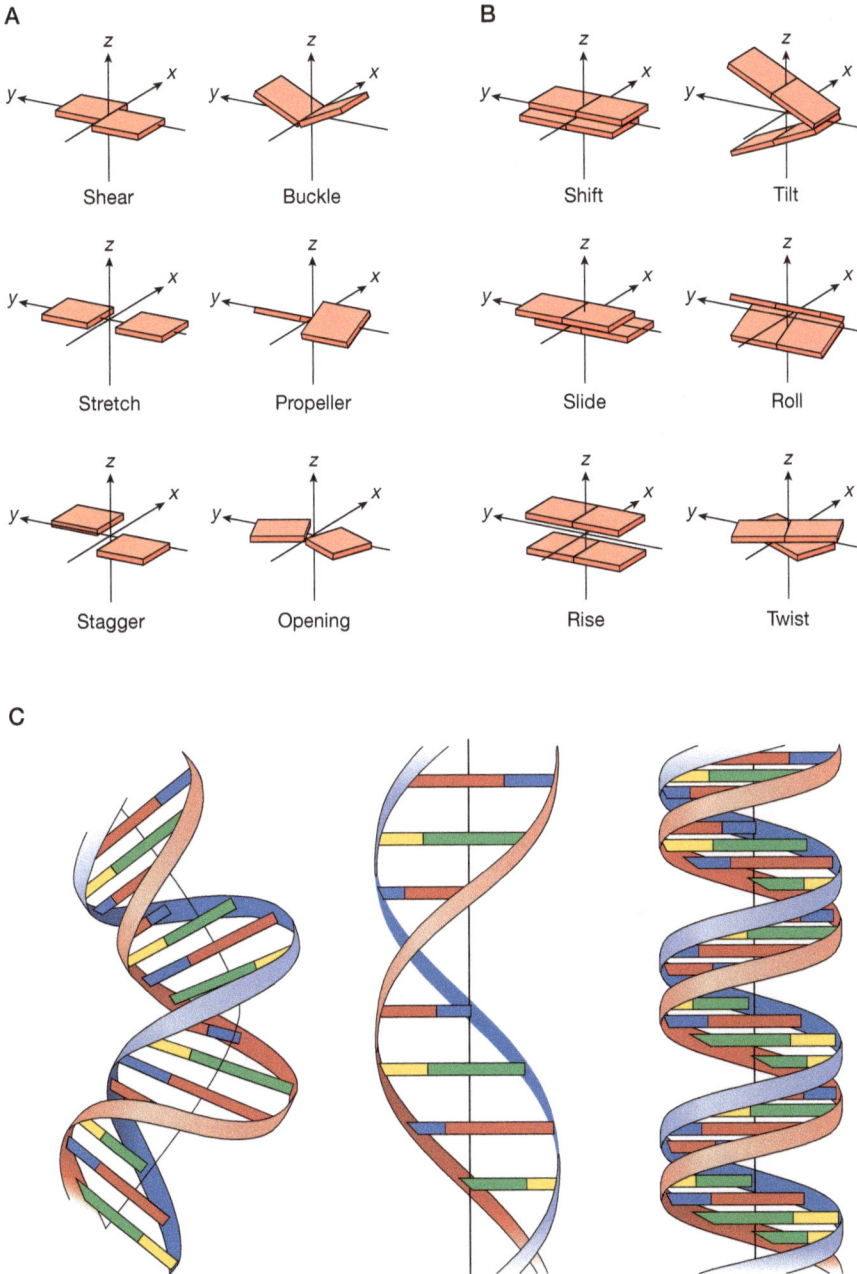

Figure 2-4. Deviations from the canonical B-form DNA structure. (*A*) Deviations of individual base pairs. (*B*) Deviations between adjacent base pairs. (*C*) The overall structure of the helix can also be deformed by bending (left), stretching (middle), and compressing (right), shown for different sections of a long helix. (*A,B* Redrawn from Lu X-J, Olson WK [2003], with permission, from Oxford University Press.)

those movements affecting the specific positioning of bases within the double-helical structure, they can also affect the overall structure of the helix itself (Fig. 2-4C). For example, the width of the major and minor grooves may become wider or narrower depending on the sequence; runs of A-T base pairs have a narrowing of the minor groove that affects their electrostatic potential and the interactions with proteins (Remo et al. 2010). The pitch of the helix may also increase or decrease. All of these movements are small, a few degrees in angle or a few angstroms (0.1 nm) in translation, but nonetheless cumulatively they can have large effects on the interactions with proteins. They can cause the DNA to form an intrinsic bend instead of the straight helical axis of the canonical structure.

Bending is essential for DNA to wrap around the histones in chromatin, and the sequence dependence of bending propensity is responsible for weak periodicities in the base composition of eukaryotic DNA. Bending can also facilitate interactions with some TFs by lowering the cost of DNA fitting the curved surface of the protein (Chapter 3). More than intrinsic bending or other deformations, the flexibility of DNA is critical and depends on the sequence. If a stretch of DNA is very rigid, it will have a large energetic cost to make the optimum contacts with a protein surface that is curved, whereas a more flexible DNA sequence can adapt to that protein's surface with much less energetic cost. The direct contacts may be between the protein and backbone of the DNA, but because the position of the backbone atoms depends on backbone shape, which depends on the sequence, this contributes to the sequence specificity of binding indirectly and is often referred to as "indirect readout." Different proteins use different combinations of direct and indirect readout, with some using direct readout exclusively and others relying on the flexibility of the backbone and the DNA structure for much of the interaction energy. The terms direct and indirect readout oversimplify the situation somewhat, and they are not mutually exclusive, but they are commonly used and provide a convenient separation of the different sources of information used by proteins to distinguish different sequences.

ALTERNATIVE DNA STRUCTURES

In addition to the variations in local structure, there can be more extensive variations. The standard form for DNA is called the B form and is represented in Figure 2-1A. Figure 2-5A compares the B form with the A form, which are similar overall, with a right-handed antiparallel helical structure, but many of the helical parameters are significantly different. For example, although the helical axis for B-DNA runs through the center of the base pairs, in A-DNA the major groove is much deeper and the helical axis is offset into the major groove. The base pairs in A-DNA are tilted $\sim 20°$ relative to the helical axis, whereas in B-DNA they are almost perfectly perpendicular. A-DNA has a pitch of nearly 11 bp but only ~ 2.7 nm, compared to 3.4 nm for the B form. It is also somewhat wider than the B form, with a diameter of ~ 2.6 nm instead of

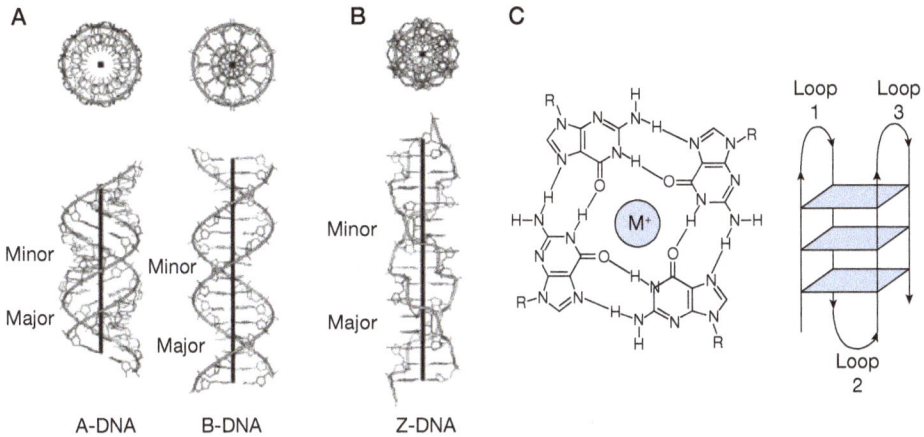

Figure 2-5. Alternative DNA structures. (A) Comparison of canonical B form with A form, in lateral views and looking down the helix. (B) Z-form DNA with a left-handed helix. (C) G-quartet structure of DNA. (A,B Redrawn from Lu X-J Olson WK [2003], with permission, from Oxford University Press; C redrawn from Wikipedia.)

2.4 nm. The A form is the normal structure adopted by dsRNA because the 2'-hydroxyl on the RNA backbone inhibits the B-form structure, and DNA can be induced into A form under appropriate conditions, such as dehydration. DNA may also adopt structures that are intermediate between B and A form, and these have been observed in some cases in interactions with proteins (Remo et al. 2010).

Another, much more different structure for DNA is also observed in vitro: Z-form DNA (Fig. 2-5B). This is preferentially formed with alternating GC runs. The structure is a left-handed antiparallel helix that is very stretched out compared to B-form DNA, the pitch being ∼4.6 nm and 12 base pairs instead of 3.4 nm and 10 base pairs for B-DNA. This form can be induced in vitro, but it is not clear if it has an in vivo role. Another unusual form of DNA that is known to have a role in vivo is the occurrence of G-quartet structures (Fig. 2-5C) that occur in ssDNA at the telomeres of chromosomes. Special proteins bind to those structures, protect the ends of chromosomes from being degraded, and facilitate their replication. But G-quartet structures are not known to be involved in gene regulation or even to occur anywhere other than at the ends of chromosomes.

One last structural variation to consider is that closed circles of DNA, or even linear DNA if the ends are constrained, can be supercoiled, either positively or negatively. This involves the helix being overwound or underwound compared to a relaxed state. Cells have enzymes—gyrases and helicases—that control the supercoiled density of DNA. Negative supercoiling can facilitate the opening up (melting) of dsDNA, thereby providing access to the individual bases. Melting of DNA is an essential part of transcription in which the RNA polymerase opens up the double helix so that one strand can be copied into the RNA. It is not clear whether TFs themselves are affected by the supercoiled status of the DNA, but certainly their downstream effects on activating or repressing transcription can be influenced. Because TFs can form loops by interacting with distant segments of the DNA simultaneously, they can constrain the DNA such that supercoiling can occur within specific segments. Again, this is controlled by cellular enzymes, but its role in gene regulation is not well understood.

For the purpose of this book, B-form DNA, with the variations described above (Fig. 2-4), is the main structure of interest. Although other structures are known to be possible, and some to exist in cells, their roles in the binding of TFs are generally unclear, and certainly thought to be of minor importance compared to the majority of interactions that have been studied.

REFERENCES

Avery OT, Macleod CM, McCarty M. 1944. Studies on the chemical nature of the substance inducing transformation of pneumococcal types. *J Exp Med* **79:** 137–158.

Beadle GW, Tatum EL. 1941. Genetic control of biochemical reactions in *Neurospora*. *Proc Natl Acad Sci* **27:** 499–506.

Britten RJ, Kohne DE. 1968. Repeated sequences in DNA. Hundreds of thousands of copies of DNA sequences have been incorporated into the genomes of higher organisms. *Science* **161:** 529–540.

Lu X-J, Olson WK. 2003. 3DNA: A software package for the analysis, rebuilding and visualization of three-dimensional nucleic acid structures. *Nucleic Acid Res* **31:** 5108–5121.

Mullis KB. 1990. The unusual origin of the polymerase chain reaction. *Sci Am* **262:** 56–65.

Olson WK, Bansal M, Burley SK, Dickerson RE, Gerstein M, Harvey SC, Heinemann U, Lu XJ, Neidle S, Shakked Z, et al. 2001. A standard reference frame for the description of nucleic acid base-pair geometry. *J Mol Biol* **313:** 229–237.

Remo R, Xiangshu J, Sean M, West R-J, Honig B, Mann RS. 2010. Origins of specificity in protein-DNA recognition. *Annu Rev Biochem* **79:** 233–269.

Watson JD, Crick FHC. 1953a. A structure for deoxyribose nucleic acid. *Nature* **171:** 737–738.

Watson JD, Crick FHC. 1953b. Genetical implications of the structure of deoxyribonucleic acid. *Nature* **171:** 964–967.

ONLINE RESOURCES

http://ndbserver.rutgers.edu Nucleic acids database. A repository of three-dimensional structural information about nucleic acids. Also contains links to educational information and other resources related to nucleic acid structures.

http://x3dna.org/ 3DNA: Suite of software programs for the analysis, rebuilding, and visualization of three-dimensional nucleic acid structures.

CHAPTER 3

Protein Structure and DNA Recognition

FROM CHAINS TO COMPLEXES

PROTEINS SHARE WITH DNA THE FEATURE OF being linear chains of residues covalently linked through a common backbone. For proteins, the residues are amino acids rather than the nucleic acid bases of DNA, and there are 20 of them instead of four. But unlike DNA, which has a nearly universal structure with slight variations, proteins have an enormous diversity of structure, determined by their three-dimensional (3D) folding patterns and the formation of multiprotein complexes (Fig. 3-1). There are four levels of protein structure. Primary structure is the linear sequence of amino acids, which is dictated by the DNA sequence of the gene that codes for the protein. Secondary structure refers to short-range structural patterns that occur commonly, predominantly the α helix and the β strand (see the section below, Secondary Structure and Tertiary Structure). Those common secondary structural elements are linked together by various turns and loops that can have a variety of shapes to generate the overall tertiary, or 3D, structure of the protein. The fourth level, the quaternary structure, refers to complexes of multiple proteins that form and are often required for function.

The sequence of a protein is determined by the DNA sequence of the gene that codes for it, although alternative splicing events can lead to the production of different proteins from the same gene. These splicing differences may be small—the addition or removal of a few amino acids—or they can be large, where entire domains exist in some versions of the protein and not in others. In addition, after synthesis modifications to proteins occur that can affect their structure and function (see Modifications, below). The secondary and tertiary protein structures are determined by primary sequence. For many proteins, the sequence contains sufficient information for proper

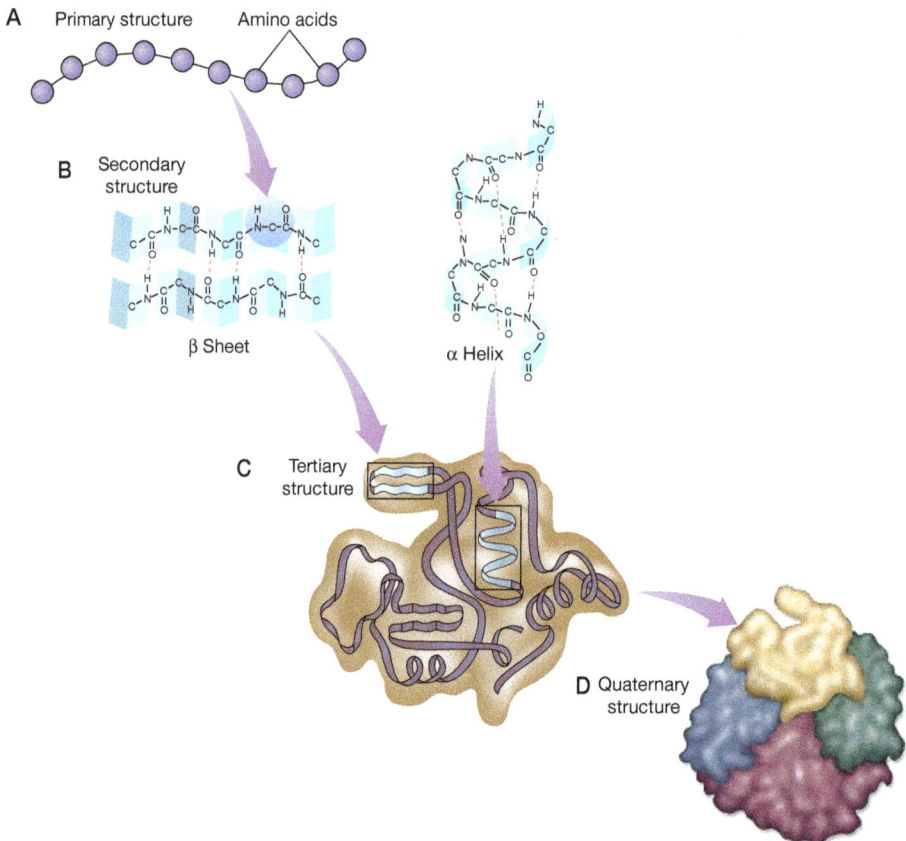

Figure 3-1. Four levels of protein structure. (*A*) Primary structure is the sequence of amino acids that comprise the linear protein chain. (*B*) Secondary structure is composed of elements such as a β sheet (*left*) and an α helix (*right*) (see Fig. 3-3). (*C*) Tertiary structure is the full 3D structure of the protein. (*D*) Quaternary structure is the 3D structure of a complex composed of several independent proteins.

folding, and these proteins can easily be denatured and renatured in vitro. Other proteins do not fold properly on their own and require "chaperones" to attain the functional tertiary structure. The tertiary structure is often not entirely rigid and can exist in two or more states or can even include domains that are largely unstructured, at least part of the time. For DNA-binding proteins, the tertiary structure can be altered by modification of the protein and by binding to small-molecule effectors that regulate their activity (see Fig. 1-1). Their quaternary structures can also be variable, such as forming homodimers or heterodimers to promote DNA binding as well as associating with auxiliary proteins that affect DNA binding and localization of the transcription factor (TF).

AMINO ACID PROPERTIES

Figure 3-2 shows the 20 amino acids of the standard genetic code that comprise the primary protein sequence. Figure 3-2A shows the backbone, composed of two carbons and one nitrogen, that is common to all amino acids. The central carbon of the backbone connects to the side chain, which distinguishes the 20 amino acids from one another. The peptide bond, linking the amino acids in the polypeptide chain, occurs between the nitrogen and the end carbon. Protein synthesis occurs in the amino-to-carboxyl direction, with the new amino acid being linked through its nitrogen atom to the end carbon atom of the previous amino acid in the chain. The protein backbone has polarity with one end (the first end synthesized), which is referred to as the amino terminus; the other end called the carboxyl terminus. A water molecule is released during peptide bond formation and the covalently linked N—H and C=O that remain are available as a hydrogen-bond donor and acceptor, respectively, and are responsible for the stable secondary structures that are commonly observed (see below).

The 20 different amino acids have a wide range of properties: They differ in size and shape, charge and polarity, preference for being surrounded by water or buried away from the solvent (hydrophilic versus hydrophobic, respectively), and many others. The specific amino acid sequence of a protein determines all of its properties (aside from those imparted by posttranslational modifications). Figure 3-2B shows the side chains of each of the 20 standard amino acids divided into five categories: nonpolar hydrophobic, polar acidic, polar uncharged, polar basic, and special cases. Each category has unique properties, although it should be pointed out that these amino acids could be divided into alternative groups using different criteria. Within each group, the amino acids are ordered from largest to smallest, top to bottom. The figure provides the name for each amino acid as well as its three-letter abbreviation and its one-letter code. For this chapter and most of this volume, the three-letter abbreviations for amino acids are used to avoid confusion with DNA bases.

Nonpolar Hydrophobic Amino Acids

Nonpolar hydrophobic amino acids have only hydrocarbon side chains with two exceptions. Met has a sulfur group in its side chain, but because this is internal and not on the end, the entire side chain behaves very similarly to a pure hydrocarbon. Trp has a nitrogen in one of its two rings, giving it a partially polar characteristic, but the large hydrocarbon rings make its behavior primarily hydrophobic. All of these side chains are more soluble in nonaqueous solutions (such as octane) than in water. This hydrophobic character means that they tend to be found buried within the protein structure away from the surface, and this tendency is one of the driving forces in the protein-folding process. Hydrophobic patches that do occur on the surface of proteins are often sites of interactions with other proteins; burying these amino acids

A

Amino acid (1) Amino acid (2)

Peptide bond

Dipeptide

Water

B

Nonpolar, hydrophobic   Polar, uncharged

R Groups

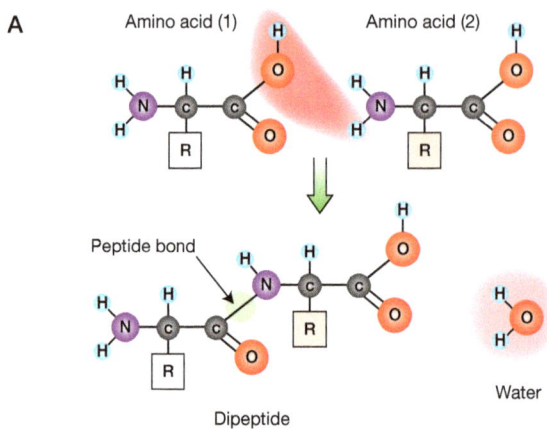

Nonpolar, hydrophobic	Polar, uncharged
Tryptophan, Trp, W, MW = 204	Tyrosine, Tyr, Y, MW = 181
Phenylalanine, Phe, F, MW = 165	Glutamine, Glu, Q, MW = 146
Methionine, Met, M, MW = 149	Asparagine, Asn, N, MW = 132
Leucine, Leu, L, MW = 131	Threonine, Thr, T, MW = 119
Isoleucine, Ile, I, MW = 131	Serine, Ser, S, MW = 105

Polar, basic

Nonpolar, hydrophobic	Polar, basic
Valine, Val, V, MW = 117	Histidine, His, H, MW = 155
Alanine, Ala, A, MW = 89	Arginine, Arg, R, MW = 174

Special cases

Special cases	Polar, basic
Cysteine, Cys, C, MW = 121	Lysine, Lys, K, MW = 146

Polar, acidic

	Polar, acidic
Proline, Pro, P, MW = 115	Glutamic acid, Glu, E, MW = 147
Glycine, Gly, G, MW = 75	Aspartic acid, Asp, D, MW = 133

Figure 3-2. (*See facing page for legend.*)

away from the aqueous environment is one of the driving forces for protein–protein interactions.

Polar Uncharged Residues

Polar uncharged amino acids all have hydrogen-bond donors and/or acceptors. Tyr, the largest, is also partially hydrophobic because of the large hydrocarbon ring (identical to phenylalanine), but the hydroxyl group on the end is a good hydrogen-bond donor. Thr and Ser also have hydroxyl groups that make them hydrogen-bond donors. Gln and Asn have both a carboxyl and an amine group, allowing them to be both hydrogen donors and acceptors. Water is an excellent hydrogen-bond donor and acceptor, and all of these amino acids are hydrophilic and tend to be located on the surface of proteins, where they can interact favorably with the surrounding aqueous environment. They can also be internal, where they often make hydrogen bonds with other polar amino acids and can help to specify the exact tertiary structure of the protein. In DNA-binding proteins, these polar amino acids frequently interact directly with the DNA because they can form favorable hydrogen bonds with both the backbone of DNA and the bases in either the major or minor groove. Because Asn and Gln can each make two hydrogen bonds simultaneously, one as an acceptor and the other as a donor, they are used frequently for DNA interactions.

Polar Basic Residues

Polar basic residues Arg and Lys are nearly always positively charged. His is often positively charged, but the association constant for H^+ is a near-neutral pH; thus, its charge depends on its local environment. These residues can form salt bridges (positive–negative interactions) with the negatively charged oxygens of the DNA backbone (see Fig. 2-2) to help stabilize the complex. They are also very good hydrogen-bond donors and can make specific contacts with the DNA bases in either the major or minor grooves. Arg, in particular, has two amino groups, can make two hydrogen bonds simultaneously (both as donors), and is frequently used for specific DNA interactions.

Figure 3-2. Chemical details of amino acids. (A) Individual amino acids have a common backbone with different side chains (R groups) that distinguish them from one another. The peptide bond links two amino acids between the amino end of one with the carboxyl end of the other, releasing one water molecule. (B) Chemical structures of each amino acid side chain, grouped in five classes based on similar characteristics (see text). Shown are the amino acid names, a three-letter abbreviation, and a one-letter code. Also listed is the molecular weight (MW) of each amino acid. Shaded area is the backbone that is common to all amino acids (except proline).

Polar Acidic Residues

Polar acidic residues Asp and Glu are nearly always negatively charged. This makes them excellent hydrogen-bond acceptors, and they are frequently used in that capacity for sequence-specific DNA interactions. Because of the negative charge on the phosphate groups of the DNA backbone, these amino acids do not make direct contact with the backbone, but those phosphate groups often interact with positively charged cations, such as Mg^{++}. Asp and Glu can interact with those cations and in so doing increase the affinity of the protein for DNA.

Special Cases

Amino acids in the special cases group have some unusual properties that set them apart. Pro uses the N of the backbone as part of its "side chain." This gives Pro less flexibility than the other amino acids and restricts the kinds of interactions that it can make. Pro rarely occurs in the middle of an α helix or β strand because of those restrictions and can help to constrain the structure of loop regions. Cys has a sulfur group on the end that can behave similarly to a hydroxyl group as a hydrogen-bond donor; the sulfur group is also commonly used in intrachain cross-linking in which two Cys residues form an S–S covalent bond to stabilize the tertiary structure of the protein. Less frequently, the sulfur group can also be used to make a covalent bond between two proteins. Gly is unique in having only an H as the side group. This makes it neither hydrophobic nor hydrophilic and gives it extra flexibility.

SECONDARY STRUCTURE AND TERTIARY STRUCTURE

Two types of secondary structures occur commonly: α helixes and β strands (Fig. 3-3). Both of these structures are stabilized by hydrogen bonding between backbone atoms, so the structures can occur in any protein sequence (although Pro occurs infrequently in these structures due to its constraints on the backbone structure).

α Helix

Figure 3-3A shows the structure of an α helix in which the carboxyl oxygen of one residue is hydrogen bonded to the amino hydrogen of the fourth residue down the chain. The pitch is 3.6 residues and ∼0.54 nm per turn of the helix. The diameter of the helical backbone is ∼0.6 nm, but the residues extend outward from the helix so that the diameter of the complete helix depends on the sequence of residues; this diameter is typically ∼1.2 nm, allowing it to easily fit into the major groove of DNA. There are two alternative possible helices: one in which the hydroxyl is hydrogen bonded to the amino group three residues further along the chain, called a 3_{10} helix, and a 5_{16} helix (π helix), where the bonding is to the amino acid five residues down

A

B

Antiparallel

Top view

Side view

Parallel

Top view

Side view

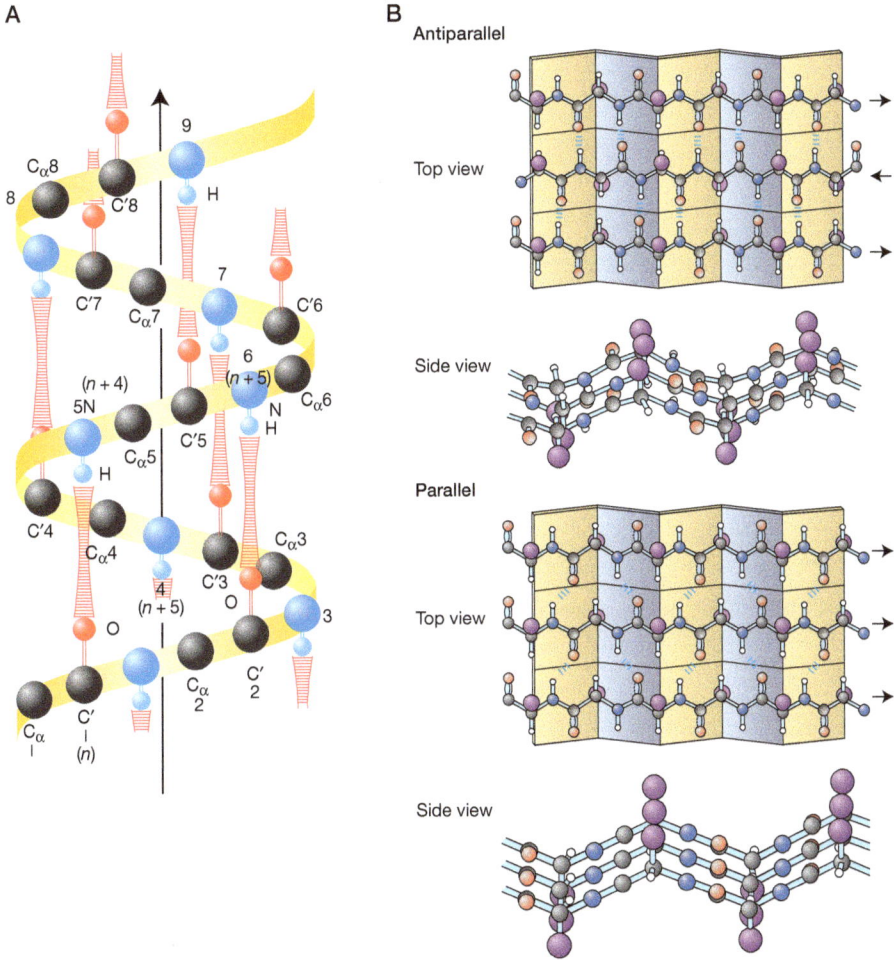

Figure 3-3. Common secondary structure elements. (A) α Helix, with hydrogen bonding between a backbone amino group and the carboxyl group of the amino acid three positions prior in the chain. (B) β-Sheet structure showing the hydrogen-bonding pattern among adjacent chains in the sheet. The individual chains may align in either a parallel or antiparallel arrangement. Side views show that side chains alternate between up and down orientations. (A, Redrawn from Niimura N. 2001, with permission from the Physical Society of Japan; B, from Lehninger Principles of Biochemistry by Nelson and Cox, ©2008, W.H. Freeman and Co. Used with permission.)

the chain. Those helices are much less common than the α helix, which is frequently used for sequence-specific DNA recognition by proteins.

β Strands and β Sheets

The other common secondary structure shown in Figure 3-3B is the β strand. In this case, extended chains occur side by side, with the backbone forming hydrogen bonds between the two chains. The chains can be either parallel or antiparallel to each other. Every second backbone hydroxyl and amino group is used to interact with a neighboring chain on one side. The alternative ones point away from that chain, so that they can be paired in the same manner to another chain, giving rise to a "sheet" (called a β sheet) of individual chains that can extend in an approximate plane for many chains in a row, with each chain either parallel or antiparallel to the others. Although typically drawn as a flat plane, as in Figure 3-3B, the β-sheet structure is not constrained to lie in a plane and can show considerable curvature. When there are only two chains involved, which is the typical case for DNA-binding interactions, it is commonly referred to as a β ribbon, with a width of ∼1 nm, which allows it to fit easily into a major groove of DNA. Although the β-ribbon structure is used much less frequently than an α helix for sequence-specific DNA recognition, some protein families use it. Note that each residue points either up or down relative to the direction of the chains and they alternate, up-down-up, etc. The adjacent residues from two chains point in the same direction. This allows the β strands to have a hydrophobic side and a hydrophilic side, which is commonly observed. The hydrophobic side can bury its side chains into the bulk of the protein, and the hydrophilic side can face the solvent and be available for interactions with other proteins or with DNA.

FUNCTIONAL DOMAINS

Complete proteins are often organized into several different domains, each performing distinct functions. Sometimes, these are completely separable, so that a segment of the protein can be isolated independently of the remaining protein and will still perform its specific function. In other cases, the entire protein, or at least a larger section of it, is needed to obtain the proper structure that is required for function. In many cases, the DNA-binding domains described in the following sections function autonomously, but the entire TF may contain additional domains that perform other functions. A TF can contain two different DNA-binding domains, although this is rare (with the exception of the zinc finger domain, which is frequently found with several such domains in tandem to create a composite DNA-binding domain that interacts with a long DNA-binding site). More commonly, as indicated in the previous section, TFs contain additional domains that allow them to dimerize, quite often as homodimers, but heterodimers are also fairly common. Additional domains can also function to specify the cellular location of the protein, such as the cytoplasm or the cell membrane, where it cannot function as a TF until modified to permit localization to the nucleus. Finally, some domains can interact with other proteins to form multiprotein complexes that regulate transcription. For example, many TFs contain an activation domain that specifically interacts with the basal transcription apparatus to increase the rate of transcription. The activation domain of the Gal4 TF is an independent domain that can be attached to other DNA-binding proteins, even bacterial repressors, to convert them into activating TFs in eukaryotes. A variety of other multiprotein complexes form on chromosomal DNA to regulate chromatin structure and transcription. The complexity of such higher-order structures is currently poorly understood.

TRANSCRIPTION FACTOR FAMILIES

Proteins can be classified into families based on their structures. Very similar structures usually have very similar functions, and TFs with similar structures usually bind to DNA with the same overall geometry of interaction. But structure determination is relatively slow experimentally, and there are only a few hundred known structures of protein–DNA complexes. Thus, most classifications are based on sequence comparisons (see Chapter 8). Proteins from the same family share a common ancestry and have diverged through evolutionary processes. Generally, they can still be recognized as being related through sequence similarity. Maintaining the proper protein structure and amino acids critical for function constrains the sequences sufficiently such that they can be recognized even over long evolutionary divergences. The following paragraphs provide brief descriptions of structural motifs of some common TF families. The structures have been determined by either X-ray crystallography or nuclear magnetic resonance (NMR) (see Box 3-1) and are available from the Protein Data Bank (PDB; see Online Resource at the end of the chapter). The structures of several examples are shown in Figures 3-4–3-8.

Helix-Turn-Helix Proteins

The first set of DNA-binding proteins to have their structures determined by X-ray crystallography contains a common structure called a helix-turn-helix (HTH) (Fig. 3-4) domain. One helix, the "recognition helix," is situated in the major groove of the DNA and contains most of the amino acids that interact directly with the DNA base pairs to provide sequence specificity. The other helix provides some interactions with the DNA backbone and helps to properly orient the recognition helix. This HTH structure was originally found in bacterial regulatory proteins, and it was initially thought that this might be a common feature of all DNA-binding proteins. In fact, most bacterial TFs, and a few eukaryotic TFs, do contain an HTH motif; however, these HTH motifs can interact with DNA in several different modes, giving rise to a variety of HTH subfamilies. These modes can differ in how the recognition helix is oriented in the major groove and some subfamilies use additional residues, outside of the recognition helix, that also contribute to sequence specificity.

Figure 3-4A shows the structure of the λ Cro protein bound to DNA, with the recognition helix highlighted in blue. Cro is a small protein, with only 61 amino acids, that binds to DNA symmetrically as a homodimer. It contains the HTH domain plus two additional helices that stabilize the structure and form the dimer interface.

Box 3-1. Structure Determination Methods

X-RAY CRYSTALLOGRAPHY

X-ray crystallography is the experimental method that has been used to determine most of the known structures for proteins, DNA, and protein–DNA complexes. Although X-ray crystallography had been used since early in the 20th century to determine the structures of small molecules, its use for determining protein structures was much more challenging. Max Perutz and John Kendrew solved the first protein structure, that of sperm whale myoglobin, in 1958 and were awarded the Nobel Prize in Chemistry for that work in 1962. The first challenge in using this method is purifying the object of interest, either the protein or DNA or both, so that the complex can be formed in vitro. Proteins can generally be expressed to high levels with current biotechnology methods but purifying active proteins can still be challenging, especially for membrane-bound proteins. Generating large amounts of any DNA sequence can be readily accomplished using either direct synthesis or polymerase chain reaction (PCR). The second challenge is obtaining high-quality crystals of the object under study, a step that is not always successful. Once crystals are obtained, they are placed in an X-ray beam where they generate a diffraction pattern (Box Fig. 3-1). The diffraction pattern is determined both by the arrangement of atoms in the crystal and the orientation of the crystal with respect to the X-ray beam; to obtain complete data about the atomic structure, diffraction patterns are collected from many different rotations of the crystal in the X-ray beam.

Spots in the diffraction pattern are determined by the spacing of atoms in the crystal, but information about the phase of the X-ray waves is lost and must be recovered to infer atomic structure. Three alternative methods are available to determine the phase. If the structure of a

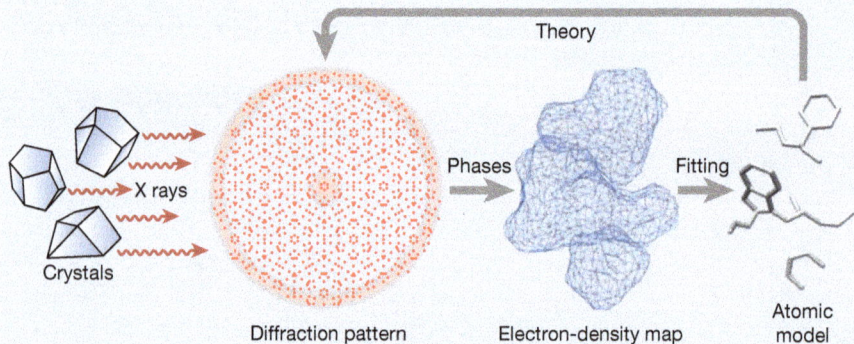

Box Figure 3-1. X rays are diffracted by the crystals and the pattern is captured. Using additional information (see text), the diffraction pattern is used to determine the electron density of the unit cell (repeating unit) of the crystal. Knowledge of the protein sequence allows amino acids to be assigned to specific regions of electron density, providing a 3D model of the protein. Given the protein structure, the diffraction pattern can be predicted, the predicted pattern can be compared to the experimental pattern, and the structure can be refined to obtain the best fit between the two.

Box 3-1. (*Continued.*)

closely related protein is known, which can be assumed to have a similar structure to the one being determined, the phases can be inferred from that previously known structure. Alternatively, one can add a small molecule, such as a metal ion, to the existing crystal, and if it goes to a particular place (or a small number of places) in the crystal, the differences in diffraction patterns with and without the added metal can be used to infer the phases of all of the spots. The third alternative is to determine the diffraction pattern using different wavelengths of X rays. Because diffraction pattern also depends on the wavelength of the incoming beam, comparisons of the patterns from multiple wavelengths can help to infer the phases of the spots.

The diffraction pattern, including phase information, is sufficient to create an electron density map, a 3D map of the atoms of the unit cell of the crystal. When this information is combined with knowledge of the protein sequence (and the DNA sequence, if the crystal contains a protein–DNA complex), it is possible to infer the overall 3D structure of the protein (or complex). The resolution of the structure (how well different atoms can be resolved) varies depending on the quality of the crystal. At 3-Å resolution, the backbone of the protein can usually be traced throughout the structure, and some information about the side-chain arrangements is visible. At resolutions of 2 Å or better, extensive information about the individual atoms of the side-chain arrangements is visible. The resolution may vary among different parts of the structure; in fact, there may be some parts of the protein that are not well resolved—essentially invisible in the diffraction pattern—because they do not have a well-defined structure and may vary among different unit cells of the crystal.

Given the 3D structure within a unit cell, it is possible to compute the expected diffraction pattern. The difference between the observed and expected patterns is referred to as the *R* factor, with 0 corresponding to an exact agreement. An *R* factor of <0.2 indicates a well-defined structure. Discrepancies between observed and expected diffraction patterns can be reduced by modifying the predicted structure, a procedure known as refinement of the structure. Typically, the initial overlay of the protein sequence onto the electron density map goes through a few refinement steps before the final, inferred structure is published and deposited in the Protein Data Bank (PDB; see Online Resource at chapter's end).

NUCLEAR MAGNETIC RESONANCE

Nuclear magnetic resonance (NMR) is a very different procedure from X-ray crystallography for determining the structure of a protein (or nucleic acid or protein–nucleic acid complex). It still requires a purified protein (and/or DNA/RNA), but it does not require a crystal. In fact, one of the advantages of NMR is that it determines the structure in solution, avoiding possible artifacts that could arise from crystal-packing forces. NMR has been used to analyze small molecules since the 1930s, but its application to proteins is much more complicated. Kurt Wüthrich received the Nobel Prize in Chemistry in 2002 for his development of methods that facilitate protein structure determination using NMR spectroscopy.

Box 3-1. (*Continued.*)

NMR is based on the fact that the nuclei of certain atoms have a magnetic moment, often referred to simply as spin. The most relevant nuclei for biological samples are ^{1}H, ^{13}C, ^{15}N, and ^{31}P (in which the superscript refers to the total number of protons and neutrons in the nucleus). ^{1}H is the most common isotope of hydrogen and occurs naturally in all proteins. The other relevant isotopes are rare and, thus, proteins must be labeled with them. For example, cells could be fed glucose that contains ^{13}C and soon all of the macromolecules in the cell will also contain ^{13}C. Or, they could be fed a specific amino acid, for example, Lys, which contains ^{13}C, and then all of the Lys in all of the proteins will be labeled with ^{13}C but other amino acids will not.

In a strong magnetic field, the individual nuclei with spin will have different energy levels depending on whether their spins are aligned parallel or antiparallel to the magnetic field. Pulses of radio frequency (RF) waves can induce nuclei to jump to higher-energy states, and this is much more efficient when RF corresponds to the energy difference of the two states, or the resonance frequency for that nucleus. The resonance frequency depends on both the type of nucleus (^{1}H, ^{13}C, etc.) and its local environment. NMR spectroscopy scans through a range of RFs and identifies those that are the resonance frequencies of nuclei within the sample. This is the one-dimensional (1D) NMR spectrum for that protein. The first challenge is to associate the resonance frequencies with specific nuclei in the protein, and that is the work that led to Kurt Wüthrich's Nobel Prize. It relies on correlations between resonance frequencies that are related to the distance between nuclei and that generate a two-dimensional (2D) NMR spectrum for the protein. COSY (correlation spectroscopy) specifically identifies nuclei that are connected through a small number (one to three) of covalent bonds (Box Fig. 3-2A). This helps to associate specific resonance frequencies with specific amino acids in the protein. NOESY (nuclear Overhauser effect spectroscopy) identifies nuclei that are near to one another in space even though they may be very distant in the protein sequence (Box Fig. 3-2B). The strength of the correlation falls off rapidly with distance, as $1/r^6$, so it is effective at identifying nuclei only up to ~5 Å apart but can distinguish the approximate distance (within that range) based on the strength of the correlation. This is critical for predicting the protein structure because it provides constraints on the distance between amino acids. Many constraints are collected typically by using a combination of ^{1}H-, ^{13}C-, and ^{15}N-labeled nuclei for proteins and also ^{31}P for nucleic acids (^{31}P can also be used for phosphorylated proteins). When sufficient constraints are collected, the structure of the protein is predicted based on computer algorithms that use both the inherent constraints of the protein sequence, such as the covalent bond lengths and propensities for specific bond angles, and the through-space constraints determined by the NOESY spectrum. Often, no unique solution fits all of the constraints optimally, so a range of structures are shown that are usually very similar but have some variation between them (Box Fig. 3-2C). This can be a result of having too few, or too noisy, constraints such that multiple different structures fit the data equally well. It can also be due to the inherent variability of the protein structure, the fact that it is dynamic in solution and not all proteins have the same structure (information that is useful to know).

Box 3-1. (*Continued.*)

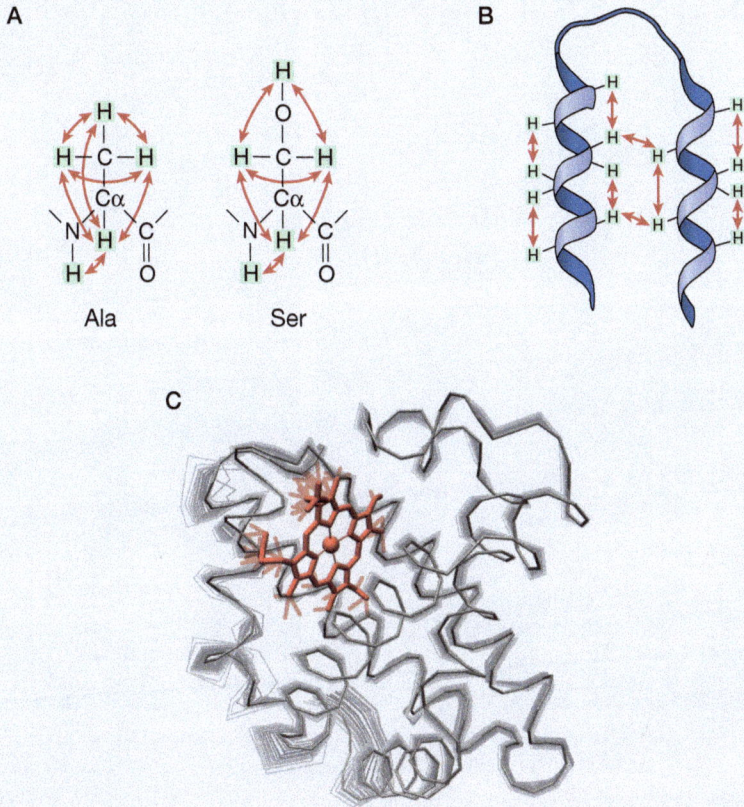

A   B

Ala   Ser

C

Box Figure 3-2. Correlation between spins can identify protons that are linked through one or two covalent bonds (*A*). The nuclear Overhauser effect (*B*) can identify protons that are close to one another in space, although they may be distant in the protein sequence. The final structure from NMR is usually a set of closely related structures that are all consistent with the experimental data (*C*), as in this NMR structure for myoglobin. (*A,B* Redrawn from Branden and Tooze, 1991; *C*, redrawn from Nelson and Cox, 2008 *Lehninger Principles in Biochemistry*, Chapter 6, Fig. 16.)

Figure 3-4B shows a dimer of the "head domain" of the Lac repressor in complex with the *lac* operator. The complete protein forms a tetramer, and each dimer can interact with a separate DNA site, leading to looping of the DNA between the two binding sites. Each monomer contains an HTH domain in which the recognition helix (blue) interacts with DNA, very similarly to other HTH domains. But this family contains

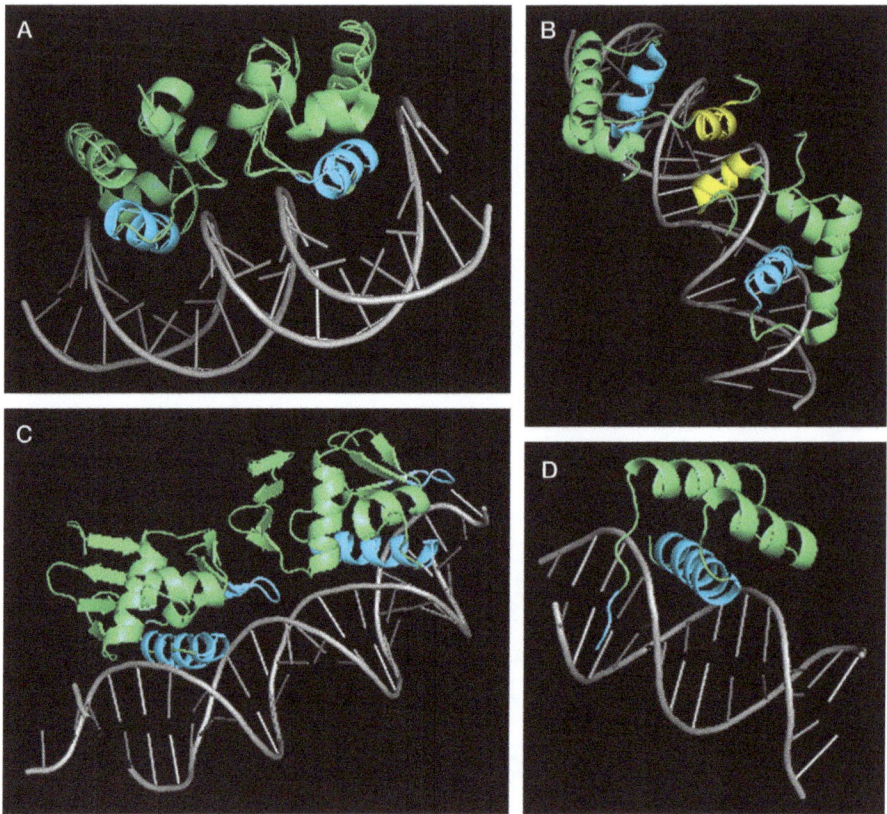

Figure 3-4. Structures of proteins with helix-turn-helix (HTH) interactions with DNA. (*A*) Cro protein dimer bound to OR1 DNA (PDB: 3CRO). (Blue) Recognition helix of each monomer. (*B*) Lac repressor protein dimer (of "head domain") bound to operator DNA (PDB: 2KEI). (Blue) Recognition helix, (yellow) additional contacts in the minor groove with a "hinge helix." (*C*) PhoB protein dimer bound to operator DNA (PDB: 1GXP). (Blue) Recognition helix and "winged" residues that make additional contacts in the minor groove. (*D*) Engrailed protein bound to consensus site DNA (PDB: 1HDD). This is a homeodomain protein. (Blue) Recognition helix and "amino-terminal arm" that makes contacts in the minor groove.

an additional helix, termed the "hinge helix" (yellow) that also interacts directly with DNA in the minor groove between the interaction sites of the two HTH domains. To accommodate both hinge helices into the minor groove, the DNA is bent away from that minor groove, which is slightly wider than normal B-form DNA.

Another variant of the HTH family is the "winged HTH" subfamily. Figure 3-4C shows the two monomers of the *Escherichia coli* PhoB protein bound to a DNA-binding site using the recognition helix of the HTH domain (blue), with the "winged" loop residues (blue) providing additional DNA contacts in the adjacent minor groove. PhoB also binds as a homodimer, but unlike the two previous proteins, this dimer is in tandem; that is, both dimers bind in the same orientation. Other members of the winged HTH family bind as symmetric homodimers, and the binding sites are approximately palindromic. In some cases, the winged loop is between the monomers, interacting in the minor groove between the two recognition helices, and in other cases the winged loop can be on the outside of the HTH domains, interacting with the minor groove further away from the center of the binding site. There are even examples of the recognition helix of the HTH domain binding in the minor groove and the winged loop providing most of the specific DNA contacts in the major groove; RFX1 is such a protein (Gajiwala et al. 2000). It binds as a symmetric homodimer to a palindromic site, with the two recognition helices binding near each other in the minor groove, reminiscent of the hinge helix region of the LacI family. But in this case, major groove interactions, which provide the majority of sequence-specific contacts, are made by the winged loop region instead of a helix.

Homeodomain proteins are a common eukaryotic TF family that uses an HTH domain as the basis of its interaction with DNA (Figure 3-4D). The recognition helix for this family is longer than most other HTH recognition helices, and a total of three helices form the homeodomain. In addition, an amino-terminal extension of the first helix interacts with DNA in the minor groove and provides additional specificity (blue). Unlike the previous HTH factors described, homeodomain proteins generally do not form dimers, although they often bind cooperatively with other TFs.

Coiled-Coil Helix Dimers

Two families of TFs contain long α helices that are used to bind specifically to DNA and also to dimerize in a coiled-coil configuration. One family has a segment rich in basic amino acids at the amino terminus of a single, long α helix; a series of spaced Leu residues at the carboxyl terminus form a "leucine zipper" at the dimer interface. This arrangement is called a basic-region leucine zipper (bZIP) motif. Figure 3-5A shows the Gcn4 protein, a member of the bZIP family that binds DNA as a homodimer, with the DNA-binding segment of the helix shown in blue. Some bZIP proteins bind DNA as heterodimers and some can form both homodimers and heterodimers. A similar family, the basic-region helix-loop-helix (bHLH) family, also has a basic region, but the dimerization domain has a loop between the two helical regions. Figure 3-5B shows a homodimer of MaxA, a member of the bHLH family, with the DNA-recognition segment shown in blue. A four-helix bundle at the interface of the loop and two helices from each monomer stabilize the interaction. Some, but not all, bHLH proteins use a leucine zipper region for a dimerization domain.

Zinc-Coordinating Proteins

Several families of TFs use zinc ions to stabilize the protein structure. The first such structure was termed a "zinc finger" because the zinc ion is coordinated between a

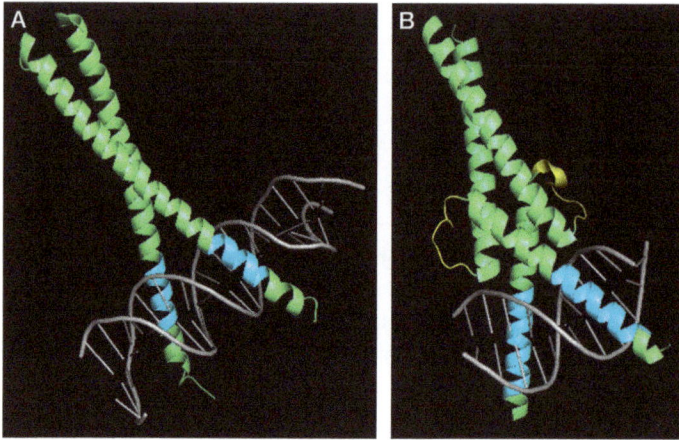

Figure 3-5. Coiled helix proteins. (*A*) Gcn4 has a basic region that interacts with DNA (blue). Long helices form dimers through a coiled-coil interaction stabilized by Leu residues that fit into the opposite helix, a "leucine zipper" structure (PDB: 1DGC). These proteins are often referred to as bZIP TFs. (*B*) Max is an HLH protein (PDB: 1HLO). Like bZIP proteins, a coiled-coil domain is for dimer formation and a basic region recognizes DNA sequences (blue). But instead of a single long helix, it has a loop region (yellow) between the two separate helical regions of each monomer. At the interface of the two helices from each monomer is a four-helix bundle that stabilizes the arrangement.

pair of β strands and an α helix, using two Cys and two His residues. Figure 3-6A shows the structure of EGR1, also known as zif268, which has three fingers, each stabilized by a zinc ion (orange), as well as DNA-recognition residues from the first six amino acids of the α helix and one residue before it (blue). Each finger contacts three base pairs, or sometimes four—including an overlap with the adjacent finger, so that the total binding site for a three-finger protein would be nine or 10 base pairs, one turn of the DNA helix. Some proteins of this type have only two fingers, but others have more than three, not all of them necessarily involved directly in DNA binding. The architecture of these TFs is fairly modular, so that different combinations of fingers can be assembled to recognize a variety of different DNA-binding sites.

The nuclear hormone receptor family of proteins also uses zinc to stabilize the structure of an α helix that interacts directly with DNA. Figure 3-6B shows a heterodimer between the retinoid X receptor (RXR) and the retinoic acid receptor (RAR) bound to DNA. In this case, there are two zinc ions (orange) for each receptor protein, and each is coordinated by four Cys residues. These proteins typically bind DNA as dimers—either homodimers or heterodimers.

In fungi, a common structure is a "zinc cluster" motif, such as that used by Gal4 (Fig. 3-6C), which uses six Cys residues to coordinate a pair of zinc atoms (orange) in each monomer of the homodimer. DNA recognition is performed by residues at the end of the α helix and in the loop following it. Gal4 prefers the interrupted palindromic sequence $CGGN_{11}CCG$, where N residues can be any bases. Other members

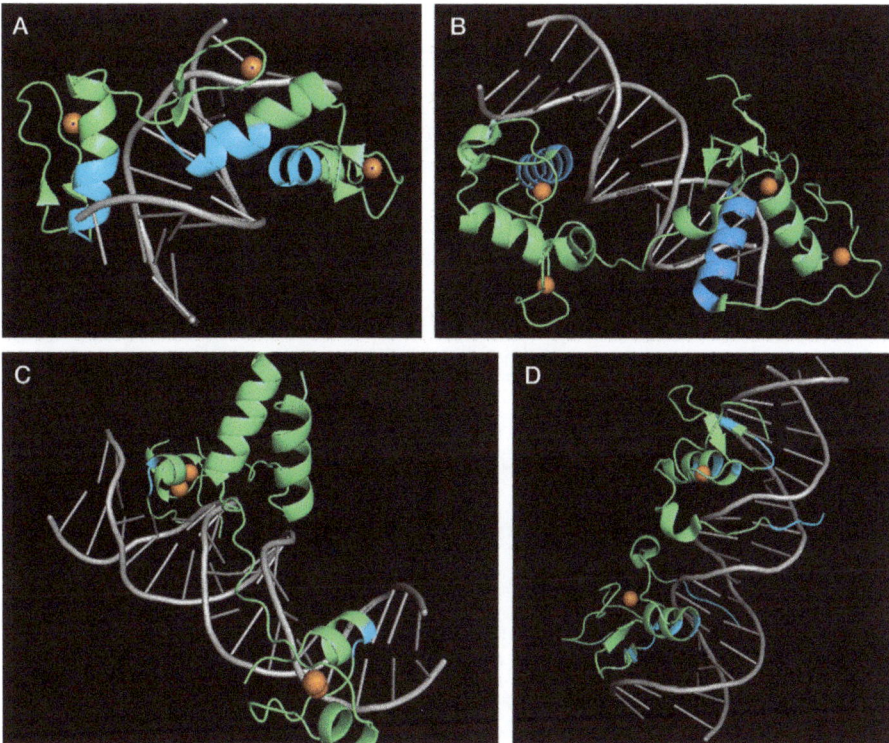

Figure 3-6. Zinc-containing TFs. (*A*) Zinc finger protein zif268 is bound to the consensus DNA sequence (EGR1) (PDB: 1AAY). This protein has three fingers, each containing a zinc ion (orange). The DNA-recognition segment of each finger includes the first six amino acids of the α helix and the amino acid just before it (blue). Each finger interacts with three base pairs so that the entire binding site is nine base pairs. (*B*) Heterodimer of the nuclear hormone receptor family proteins RXR and RAR bound to DNA (PDB: 1DSZ). Each monomer coordinates two zinc atoms (orange) and inserts a recognition helix into the major groove (blue). (*C*) Gal4 homodimer bound to consensus DNA (PDB: 1D66). Each zinc-containing helix interacts with a three-base-pair sequence of CCG, symmetrically related to one another, with a spacing of 11 base pairs between them. Other members of this family have different spacings between the interacting bases. (*D*) Two GATA factor proteins bound to two adjacent DNA sites (PDB: 3dfv). DNA base interactions (blue) come from the helices stabilized by the zinc atoms (orange) as well as additional amino acids in the β-strand segment and some loop residues.

of this family also bind to CGG sequences but with different spacings between them and sometimes in direct, rather than inverted, orientations.

A fourth family of zinc-coordinating TFs is the GATA factors, so named because they bind to the DNA sequence GATA (Fig. 3-6D). Their overall structure is somewhat similar to zinc finger proteins, but here, the zinc ion (orange) is coordinated by four Cys residues, and DNA-recognition residues (blue) come from some residues within the α helix as well as the β strand and a carboxy-terminal arm that interacts in the minor groove. Each GATA factor contains two such fingers, but often only one is used for binding, whereas in other cases, both are used. When only one is used, the factor often binds as a dimer—either homodimer or heterodimer—with other GATA factors.

Recognition with β Strands

As indicated in the above examples, most DNA-binding proteins use α helices for sequence-specific recognition, often accompanied by additional residues. But in addition, TF families use β strands exclusively for recognition. One striking example is the TATA-binding protein TBP (Fig. 3-7A), which uses a 10-strand β sheet to bind the minor groove of DNA that is grossly distorted from normal B-form structure. The minor groove is widened to the point of being nearly flat, and the DNA is bent away from the minor groove into a U shape. Normal B-form DNA can also be bound by a two-strand β ribbon, as in the case of the MetJ protein (Fig. 3-7B). Here, the protein binds as a homodimer, with each monomer contributing one β

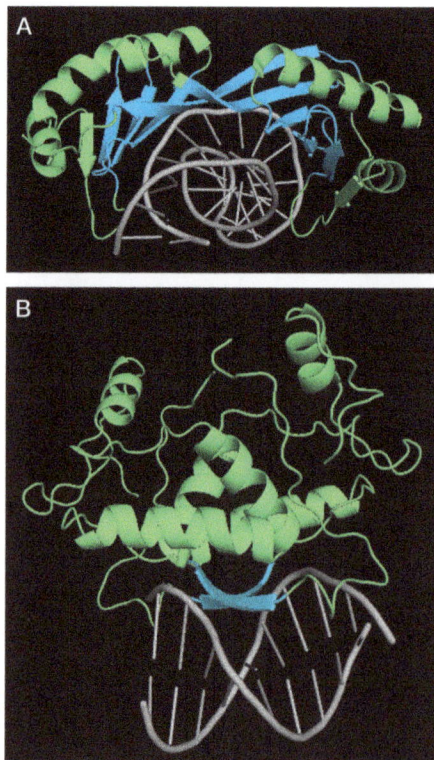

Figure 3-7. DNA recognition with β-sheet domains. (A) TATA-binding protein (TBP) bound to TATA box containing DNA (PDB: 1YTB). (Blue) Eleven-stranded β sheet that interacts with minor groove of DNA. (B) MetJ protein dimer bound to consensus DNA (PDB:1CMA). (Blue) β Ribbon composed of one strand from each monomer.

strand to the antiparallel β ribbon; the binding site is approximately palindromic. Other members of the ribbon-helix-helix family, such as Mnt and Arc, bind as homo-tetramers, each dimer of which has a structure similar to that of MetJ, with the β ribbon being formed from the amino terminus of each monomer.

Recognition with Loops

Finally, there are TFs that fold into a structure similar to an immunoglobulin domain, often referred to as the Rel-homology domain based on the structure of the c-Rel oncogene bound to DNA (Fig. 3-8A). Other TFs that bind DNA similarly include p53, runt, the STAT factors, NF-κB, and nuclear factor of activated T cells (NFAT). The DNA-recognition residues occur in loops (blue) connecting β strands from the immunoglobulin-like domain, and sometimes also α-helical segments, and they often bind as dimers.

The structure of the protein–DNA complex of transcription activator-like (TAL) effectors, bacterial proteins that are injected into plants where they function as TFs, has recently been determined (Deng et al. 2012; Mak et al. 2012). The proteins bind DNA using multiple TAL repeats, each 33 to 35 amino acids in length and composed of a two-helix bundle. A short loop connecting the two helices inserts into the major groove of DNA where one or two of the amino acids in the loop make specific contacts to a single base pair. The entire TAL effector protein has many of those repeats, 23 in PthXo1 (Mak et al. 2012) and 11 in dHax3 (Deng et al. 2012), allowing them to bind specifically to quite long binding sites. There are undoubtedly additional TF families yet to be discovered.

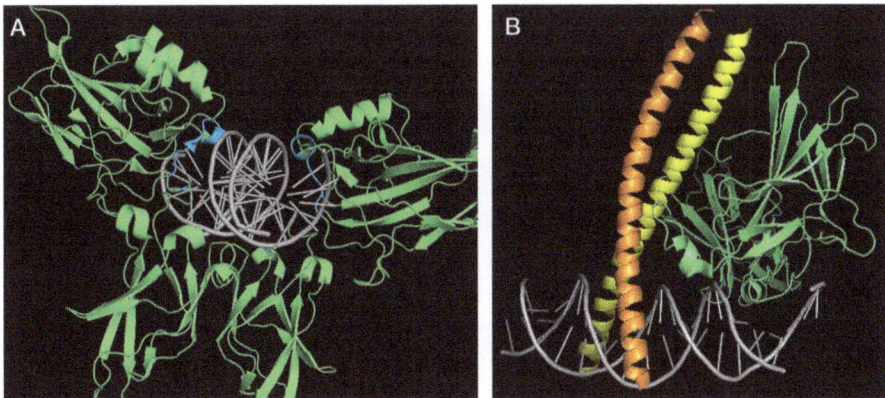

Figure 3-8. (A) Structure of the c-Rel oncogene bound as a dimer to DNA (PDB: 1gji). Recognition residues (blue) occur in the loops connecting several strands of a domain that is similar to a fold common in immunological proteins used for antibody binding. (B) NFAT–Jun–Fos complex (PDB: 1A02). Complex of a rel-homolog NFAT (green) bound to DNA adjacent to a heterodimer of two bZIP proteins: Fos (yellow) and Jun (orange). Multiple contacts between the bZIP proteins and NFAT protein facilitate binding all of them to adjacent sites on the DNA.

MULTIPROTEIN COMPLEXES

Many TFs work in combination rather than individually. The preceding sections presented many examples of proteins binding as homodimers and also a few as heterodimers, both of which are quite common. The size of the genome, especially eukaryotic genomes, and the relatively short length of the binding sites for individual proteins require combinatorial interactions to achieve control of gene expression, which is covered in more detail in Chapter 9. Figure 3-8B shows one example in which the details of the interactions have been determined by crystallography. The transcription factor NFAT (green), a member of the Rel-homology family, binds adjacent to a bZIP heterodimer containing Fos (yellow) and Jun (orange). The zipper region of the bZIP dimer is bent toward NFAT to create multiple contacts between the proteins, contributing to the binding affinity of the entire complex.

Several other structures showing combinatorial binding have been determined but mostly for complexes involving two or three proteins. More complex structures that include many more proteins, both TFs and auxiliary proteins that contribute to the regulation of gene expression, are likely to exist but are largely uncharacterized at the present time.

MODIFICATIONS

Proteins can be modified in various ways that affect their function. One significant type of modification that is irreversible is protein cleavage. For example, the p50 subunit of the p50/p65 heterodimeric TF NF-κB/RelA is synthesized as a larger p105 precursor containing an ankyrin repeat domain that localizes the protein to the cytoplasm. After cleavage of the ankyrin domain, the p50 domain joins with p65 to form a heterodimer that translocates to the nucleus, where it functions as a TF.

Other modifications are less severe and often reversible. Histones are modified with methyl and acetyl groups on various Lys residues, and those modifications constitute "epigenetic marks" that affect chromatin structure and accessibility of TFs to specific genomic regions.

The most frequent modification to TFs is phosphorylation. Ser and Thr are commonly phosphorylated on their hydroxyl group, and Tyr can also be phosphorylated. These phosphorylations are often the end product of a signaling cascade in which an initial signal causes phosphate groups to be added to proteins that in turn pass them on to other proteins. Phosphorylation of TFs can affect their overall structure, modifying their ability to bind DNA specifically; it can also affect their interactions with other proteins and might allow them to enter the nucleus or alter their stability.

A common signaling pathway in bacteria is the two-component system, in which a membrane-bound histidine kinase's activity is modulated in response to an environmental signal. When that signal is activated, such as by binding to an appropriate ligand in the cell's environment, the kinase passes the phosphate group to an Asp residue on a response regulator. The response protein is often a TF, and the phosphorylation event alters the structure to activate it for DNA binding. By such signaling pathways, the cell can alter its transcriptional profile in response to the chemical environment.

ALLOSTERIC EFFECTORS

In addition to modifying the activity of TFs by protein modification, the function of many TFs is influenced by binding to small-molecule effectors. The binding of these effectors typically alters the structure of the DNA-binding domain to either increase or decrease its affinity and/or specificity for DNA. For example, the *lac* repressor binds to the *lac* operator to inhibit synthesis of genes required for lactose metabolism that are not needed and would be of no use to synthesize if no lactose is available to the cell. If lactose is present, it binds to the repressor and reduces its affinity for the operator thereby leading to expression of lactose-metabolizing genes. Another example is the Trp repressor, which controls the synthesis of genes needed to make tryptophan. If a sufficient supply of tryptophan is present in the cell, there is no need to make those enzymes, so the repressor binds DNA to inhibit their expression. In this case, the small-molecule Trp changes the structure of the factor so that it binds the operator with higher affinity. Numerous other examples exist in which small molecules can bind to TFs to regulate their binding activity. This is a general means of feedback control whereby the genes that are regulated by the TF are only expressed under conditions in which they are useful to the cell.

Chapter 2 describes how the DNA sequence contains information for the synthesis of proteins as well as for the regulation of their expression. Proteins can distinguish among different sequences through complementarity between chemical groups on the edges of base pairs and amino acid side chains as well as through variations in DNA structure that conform to the protein's surface. Multiple different protein architectures can be used to read the DNA sequence. Often, an α helix is positioned in the major groove to interact with base pairs, but there are a variety of distinct geometries between DNA and protein. In Chapter 4, we peer into the interface between protein and DNA to see how specific contacts provide the sequence-specific discrimination ability required for regulatory proteins to function.

REFERENCES

Deng D, Yan C, Pan X, Mahfouz M, Wang J, Zhu JK, Shi Y, Yan N. 2012. Structural basis for sequence-specific recognition of DNA by TAL effectors. *Science* **335:** 720–723.

Gajiwala KS, Chen H, Cornille F, Roques BP, Reith W, Mach B, Burley SK. 2000. Structure of the winged-helix protein hRFX1 reveals a new mode of DNA binding. *Nature* **403:** 916–921.

Mak AN, Bradley P, Cernadas RA, Bogdanove AJ, Stoddard BL. 2012. The crystal structure of TAL effector PthXo1 bound to its DNA target. *Science* **335**: 716–719.

Niimura N. 2001. Neutron protein crystallography in JAERI. *J Phys Soc Jpn* **70**: 396.

FURTHER READING

Branden C, Tooze J. 1991. *Introduction to protein structure*. Garland Publishing, New York.

   An excellent overview of protein structure and methods for structural analysis.

There are many valuable review articles about protein–DNA interactions. This following list includes those on which we relied for the information presented in this chapter.

Akinori S, Hidetoshi K. 2005. Protein-DNA recognition patterns and predictions. *Annu Rev Biophys Biomol Struct 2005* **34**: 379–398.

Gajiwala KS, Burley SK. 2000. Winged helix proteins. *Curr Opin Struc Biol* **10**: 110–116.

Garvie CW, Wolberger C. 2001. Recognition of specific DNA sequences. *Mol Cell* **8**: 937–946.

Luscombe NM, Austin SE, Berman HM, Thornton JM. 2000. An overview of the structures of protein-DNA complexes. *Genome Biol* **1**: S001.

Pabo CO, Sauer RT. 1992. Transcription factors: Structural families and principles of DNA recognition. *Annu Rev Biochem* **61**: 1053–1095.

Rohs R, Jin X, West SM, Joshi R, Honig B, Mann RS. 2010. Protein-DNA recognition. *Annu Rev Biochem* **79**: 233–269.

Sarai A, Kono H. 2005. Protein-DNA recognitions patterns and predictions. *Annu Rev Biophys Biomol Struct* **34**: 379–398.

Wolberger C. 1999. Multiprotein-DNA complexes in transcriptional regulation. *Annu Rev Biophys Biomol Struct* **28**: 29–56.

ONLINE RESOURCE

http://www.rcsb.org/pdb   Portal to the Research Collaboratory for Structural Bioinformatics Protein Data Bank. This database contains structural information for proteins, nucleic acids, and protein–nucleic acid complexes determined by X-ray crystallography, NMR spectroscopy, and cryoelectron microscopy. Each entry contains a variety of information about the protein(s) and/or nucleic acid(s) whose structure has been determined, including the coordinates of each atom that can be obtained from the structural information (in a defined coordinate system). The portal also contains a variety of other information, including tutorials, related to macromolecular structures. A variety of software tools are available to visualize the structures available in the PDB. The structures shown in Figures 3-4 to 3-8 were rendered using PyMol software (http://www.pymol.org/). For each structure, the PDB identifier is given (PDB: xxx, where xxx refers to the entry in the PDB).

CHAPTER 4

Sequence-Specific Interactions in Protein–DNA Complexes

THE PREVIOUS CHAPTER SHOWED THE OVERALL GEOMETRY OF protein–DNA complexes and how the surface of the protein can accommodate the DNA double helix, sometimes bending or distorting the DNA to increase the interaction surface and the number of favorable contacts. This chapter examines the interface between protein and DNA to determine how sequence specificity is obtained—that is, the details of how the protein has different affinities for different DNA sequences. Later chapters describe in more depth how specificity is measured and modeled, but this chapter focuses on structural details that indicate how complementarity between protein side chains and DNA base pairs contributes to specificity.

It is worth noting that there are two related but distinct meanings commonly applied to the term "specificity." One is the difference in binding affinity between a preferred binding site and bulk DNA, usually referred to as the nonspecific binding affinity of the protein. Some proteins are referred to as nonspecific binders because they bind to all sequences with essentially the same affinity. In contrast, sequence-specific binding proteins usually bind with the highest affinity to a single sequence (sometimes there is a small set of sequences with equivalently high affinity), and this binding affinity is typically 10^3- to 10^6-fold higher than affinity to overall, random DNA. This is the maximum specificity of the protein—the difference between its highest- and lowest-affinity sites.

The second common meaning of specificity is related to how affinity is distributed across sequences. Some proteins, such as restriction enzymes that cut DNA at specific sequences, are essentially active in an "all-or-none" fashion; they cut only at perfect matches to their restriction site sequence. Those enzymes show high specificity on a very fine scale, in which single variants from the restriction site sequence are essentially inactive. Most transcription factors (TFs) do not have such a high specificity at the scale of single-base changes. They usually have a preferred sequence, which has the highest affinity of any sequence, but other sequences that differ in only one position often have similar affinities. Sites with multiple changes from the preferred sequence may even have a reasonably high affinity in some cases. In general, the distribution of binding affinities for TFs is much closer to a continuum than the highly discrete distribution of restriction enzymes. In such cases, measures such as "information content" (see Chapter 7) that take into account the entire distribution of binding affinities are more useful than just the ratio between the highest- and lowest-affinity sites.

There are simple rationales for the two different types of specificity. The restriction enzyme serves to protect the cell from invading DNA sequences, such as viruses, by cutting any DNA that contains its restriction site sequence. It protects its own DNA by modifying those sequences, usually with methylases that recognize the same site, but if the enzyme were less specific and cut at many different sequences, it would be susceptible to self-destruction. TFs, on the other hand, typically regulate many different genes, and it may be beneficial to the cell to have those genes expressed at different levels. By having the same TF bind to some sites more strongly than to others, which translates into a higher probability of binding or higher average occupancy, this can regulate the genes to different levels using the same machinery. Of course, this is only one part of the mechanism of differential gene regulation, but the variability in binding affinity due to reduced fine-scale specificity of TFs, compared with restriction enzymes, contributes to the fine-tuning of gene expression through the altering of regulatory DNA sequences.

SOURCE OF SPECIFICITY: THE PROTEIN–DNA INTERFACE

Interactions between proteins and DNA that provide specificity are often divided into two classes: (1) Direct readout refers to contacts, both hydrogen bonds and van der Waals interactions, between the protein and the accessible edges of the pairs; (2) indirect readout refers to contacts, primarily to the backbone, that depend on the specific structure of the DNA, such as being bent or the grooves being wider or narrower than standard B-form DNA. Direct interaction with base-pair edges allows discrimination between sequences because of the different patterns of hydrogen-bond donors and acceptors as well as the methyl group on T (see Fig. 2-2B,C). Although not dependent on direct contacts with base pairs, indirect readout also provides sequence specificity because different deformations of the DNA structure (see Fig. 2-4) are also sequence dependent, with some sequences being more likely (energetically more favorable) to adopt certain structures than other sequences. It is also possible for a base pair to contribute direct interactions with the protein and to contribute through an influence on the local DNA structure, suggesting that classifying base-pair contributions into mutually exclusive categories of direct and indirect is insufficient. A more general classification includes direct base-pair contributions from the contacts between their edges and the protein, and contributions from the shape of the DNA, which facilitates additional favorable interactions, including the DNA backbone, and that individual base pairs may be in both classes simultaneously (Rohs et al. 2009).

Most TF families described in Chapter 3 interact with double-stranded DNA and primarily in the major groove (see Figs. 3-4 through 3-8). Many also have additional contacts in the minor groove, such as those of several members of the helix-turn-helix (HTH) family (see Figs. 3-4B through 3-4D). Interactions between amino acid residues and base pairs in both the major and minor grooves provide direct readout of the DNA sequence. This is governed by different amino acids preferring different base pairs due to complementarity between hydrogen-bond donors and acceptors as well as van der Waals interactions. In addition, some of the DNA-binding sites are bent to make more contacts with the DNA than would perfectly straight DNA. The flexibility of the DNA, allowing it to bend without too much energetic cost, is also determined by DNA sequence (Chapter 2) but does not involve direct contacts between protein and DNA base pairs and contributes as indirect readout. In addition, many contacts between protein and DNA backbone are not dependent on any distortions from standard B-form DNA. These contacts provide affinity of protein for DNA but do not discriminate among different DNA sequences and therefore do not contribute to specificity (see Chapters 5 and 6).

Many proteins bind to DNA nonspecifically, where every sequence has essentially the same affinity for the protein. It is easy to imagine how that can be accomplished because the DNA backbone (see Fig. 2-2A) is highly negatively charged; thus, a protein with appropriately spaced positively charged amino acids can form other electrostatic interactions directly with the backbone, either to one strand alone or to both strands simultaneously. But even for most sequence-specific DNA-binding proteins, the majority of contacts between protein and DNA are nonspecific contacts with backbone atoms. They provide a large component of the total binding energy and allow the proteins to bind in both specific and nonspecific binding modes (see below). Nonspecific binding modes allow proteins to "slide" along DNA until they find their high-affinity binding sites, facilitating the search for functional binding sites (see Chapter 6). When the protein encounters a high-affinity DNA sequence, it can alter its orientation relative to the DNA, with increased affinity relative to nonspecific binding mode (see below).

Although the majority of specific contacts with DNA are through an α helix of the DNA-binding domain, the geometry of the interaction is highly variable (Garvie and Wolberger 2001). Sometimes, the entire helix fits into the groove parallel to the backbones of the two strands; this orientation affords the greatest number of contacts. But in other cases, only a portion of the helix is inserted in the groove, and it can be either the amino- or carboxy-terminal portion of the helix. The orientation of the helix with respect to the backbones is variable, in some cases affording only a couple of direct interactions between amino acids in the helix and DNA base pairs. Within a family or a subfamily (such as for the HTH family, which has many subfamilies with specific features), orientations are quite consistent and are determined largely by how the protein interacts with the DNA backbone to orient the recognition helix into the major groove to govern direct readout. The minor groove is normally much narrower and α helices that do interact there do not really fit into it; they just contact it tangentially and usually only for a few base pairs unless it is widened, such as by binding to the Lac repressor protein (see Fig. 3-4B). An exception to the general rule of most contacts being in the major groove is the TATA-binding protein of eukaryotes (see Fig. 3-7A), which binds directly and exclusively in the minor groove using a β sheet composed of multiple strands. To accomplish this interaction, DNA is bent dramatically, the minor groove is widened and flattened, and protein and DNA make multiple direct interactions.

Several years ago, in a compilation of >120 crystal structures of protein–DNA complexes, Luscombe et al. (2001) observed that every polar amino acid could be found forming a hydrogen bond with a base pair in at least one structure, and most of them were used by many different proteins. Furthermore, most of the amino acids were observed to form hydrogen bonds with more than one base; in many cases, there were examples of all four different bases interacting with the same amino acid in different structures. There were still obvious preferences; for example, Arg hydrogen bonding with G was by far the most common interaction, often forming two hydrogen bonds between hydrogen-bond donors of Arg (see Fig. 3-2B) and two hydrogen-bond acceptors of G (see Fig. 2-2C). Gln and Asn can also form two hydrogen bonds with A, an interaction that is commonly observed. Essentially, every hydrogen-bond donor from amino acids (Fig. 3-2B) can be found interacting with every hydrogen-bond acceptor available in the major groove (Fig. 2-2B,C) in at least one solved structure. The same is true for hydrogen-bond acceptors on protein and donors on base pairs; that is, there are many ways in which hydrogen bonds can be formed between amino acids and base pairs, and essentially all of these occur, some more frequently than others. Because the protein backbone is also capable of forming hydrogen bonds, in some cases it is used to interact directly with a DNA base, in which case the amino acid can even be a nonpolar one. There are also many direct contacts with the base pairs using van der Waals contacts with amino acid side chains. This is particularly prevalent with the methyl group of T, which sits in the major groove and provides easy access to hydrophobic groups on the amino acids; but other bases can also form van der Waals contacts. The large variety of interactions allowed between amino acids and base pairs, as well as rearrangements at the interacting protein surface (see below), makes it clear why there is no simple code relating the sequence of a protein to its preferred binding site.

PROFILES OF SPECIFICITY

Several examples illustrate how structural analyses can illuminate the mechanisms of specificity. These primarily focus on direct readout mechanisms in which specific hydrogen-bond and van der Waals contacts between amino acids and base-pair edges can be inferred from structures of protein–DNA complexes. The first is the restriction enzyme EcoRI, which shows how very high specificity at the base-pair level can be obtained. The other examples are from three different TF families, all of which tolerate variations in their binding sites with relatively minor decreases in binding affinity. Many structures of protein–DNA complexes are now available in the Protein Data

Bank (PDB; see Chapter 3); however, there are only a few for which the structures have been solved for the same protein bound to different DNA sequences with measured changes in binding affinity. Examining a few of these cases illustrates plasticity at the protein–DNA interface, where rearrangements of the amino acids can compensate for loss of some important contacts. These characteristics of TFs allow them to achieve high maximum specificity, in some cases $>10^6$-fold, while at the same time allowing a fairly gradual loss of affinity over many different variations—essentially, a continuum from highest affinity down to nonspecific binding.

EcoRI

The restriction enzyme EcoRI binds and cuts the sequence GAATTC. *Escherichia coli* also makes the enzyme EcoRI methyltransferase, which methylates the second A (on both strands of the palindromic site) to protect those sites in its own genome from being cut—a possibly lethal event. The enzyme cuts both strands between G and A, leaving a double-stranded break in the DNA. The extreme selection for cutting only at GAATTC sites is not due solely to selection in binding; EcoRI will bind to other sequences at low probability and sometimes even cut one strand, which is easily repaired. But studies of binding alone have shown a very high specificity, even for alternative sites with only one mismatch. Figure 4-1 shows some of the contributions to the specificity. Not shown are the many contacts to the backbone or kinks in the DNA at the G-A cleavage site and also in the middle of the binding site. Those distortions are facilitated by DNA sequence and represent shape contributions to specificity. Also not shown are all of the interactions between the protein residues, which

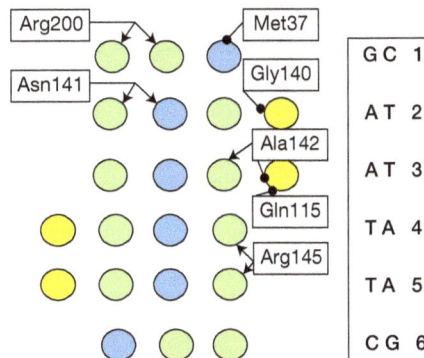

Figure 4-1. Interface of the EcoRI restriction enzyme with its restriction site showing primary contacts in the major groove (Pingoud and Jeltsch 1997). The protein binds as a dimer and only one of two monomers is shown; the other interacts in exactly the same way in the opposite orientation of this palindromic site. The sequence is shown in the box at right. Bases are displayed using colors in Figure 2-2B,C: (Blue) Hydrogen-bond donors, (green) hydrogen-bond acceptors, and (yellow) the methyl group on T. (Arrows) Hydrogen bonds, (small black circles) van der Waals interactions.

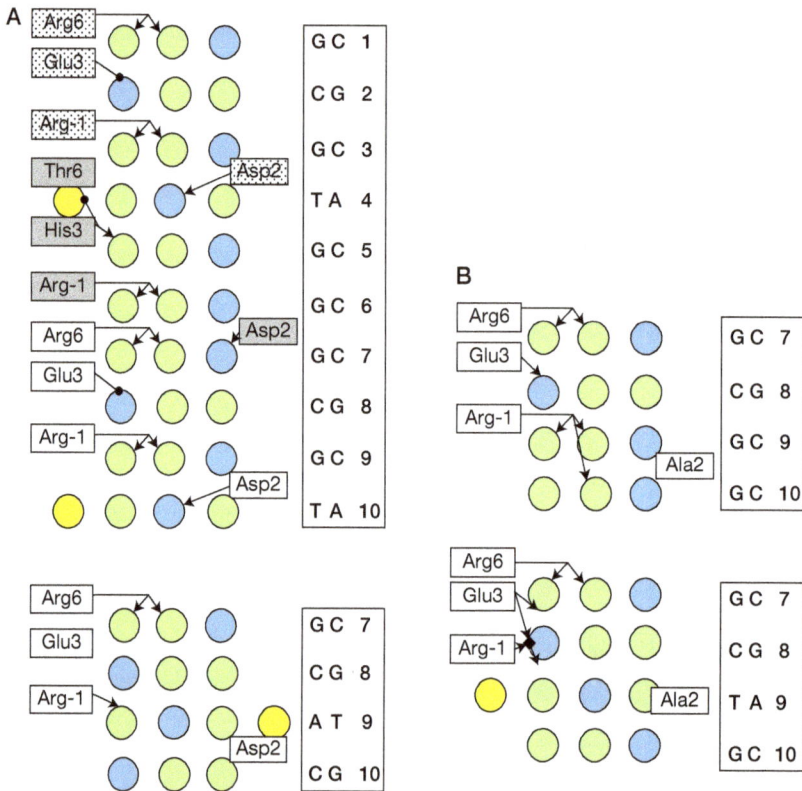

Figure 4-2. Interactions of zinc finger proteins with different DNA sequences (Elrod-Erickson et al. 1998; Miller and Pabo 2001). (*A, top*) Wild-type zif268 protein interaction with its consensus binding site GCGTGGGCGT. For each of the three fingers, amino acids at positions −1 (relative to the α helix), 2, 3, and 6 are shown. These amino acids make critical contacts with DNA in this "canonical" interaction between a zinc finger protein and DNA. Wild-type zif268 protein has critical residues RDER in positions −1, 2, 3, and 6 of finger 1 (and finger 3). The other proteins, shown below, have the same finger 2 and 3 sequences but vary in finger 1; thus, only this interaction is shown. (*A, bottom*) Wild-type zif268 bound to an alternative binding site with GCAC at positions 7–10. (*B, top*) zif268 mutant D20A has critical residues RAER in finger 1, bound to the wild-type consensus sequence. (*B, bottom*) D20A bound to the alternative binding site with sequence GCTG in positions 7–10. (*Legend continues on following page.*)

make the interacting residues quite rigid and unable to be rearranged easily and adjust to different sequences. The figure shows the many direct contacts, both hydrogen bonds and van der Waals interactions, between protein and the specific sequence of the restriction site. If any base pair were to be replaced by any other, multiple favorable contacts would be lost, and because the protein is very rigid and cannot easily compensate for those lost contacts through rearrangement, this would lead to a large decrease in binding affinity.

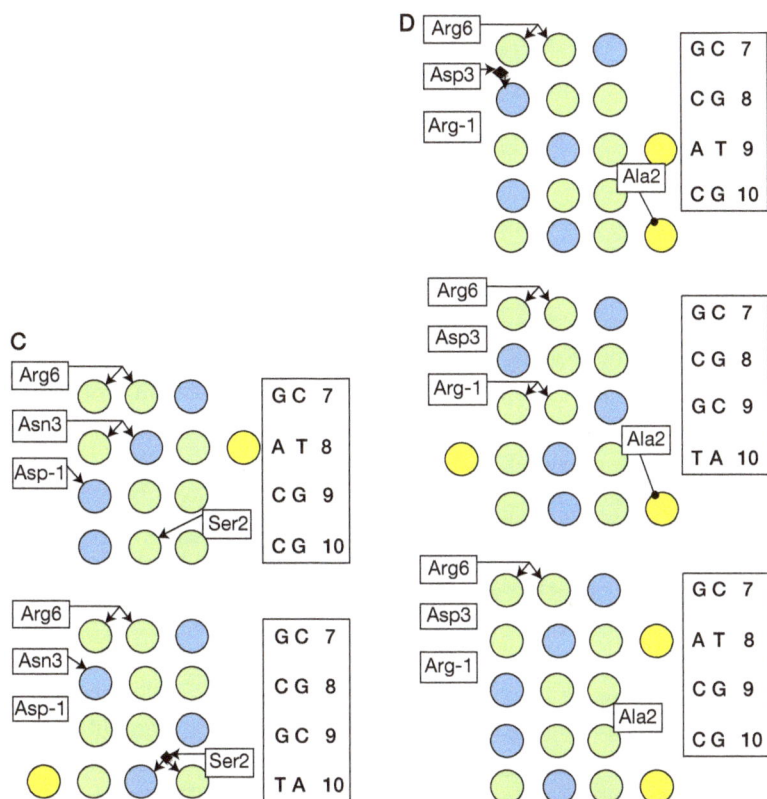

Figure 4-2. *(Continued.)* *(C, top)* Variant of zif268 with critical residues DSNR in finger 1 bound to the sequence GACC. *(C, bottom)* DSNR bound to wild-type sequence GCGT. *(D, top)* zif268 variant with critical residues RADR bound to sequence GCAC. *(D, middle)* RADR bound to wild-type binding site GCGT. *(D, bottom)* RADR bound to alternative sequence GACC (black diamonds are bound water molecules).

Zinc Finger Proteins

Zinc finger proteins are undoubtedly the most studied of all TF families. An individual finger is composed of an α helix and two β strands that coordinate a zinc ion and position the α helix into the major groove of DNA, where it can interact directly with base pairs. Each finger typically interacts with three to five base pairs, and multiple fingers can be concatenated in the protein to allow specific binding of longer sequences (see Fig. 3-6A). The first structure was determined for the protein zif268, also known as EGR1, and it has been the basis for many other studies. Because many other proteins bind DNA using the same positions in the α helix interacting with the same positions in the binding site, that particular interface is often referred to as the canonical zinc finger–DNA interaction (Fig. 4-2A), although other interaction arrangements are seen in other zinc finger proteins. zif268 has been the basis for many studies in which proteins have been selected from randomized libraries that bind to specific DNA sequences, and also the converse, in which DNA sequences have been selected that bind to specific proteins (see Chapters 5 and 8).

There are some general rules about preferred interactions—which amino acids at which positions prefer to bind to which bases at the interacting positions—and some algorithms have been developed to predict the specificities of particular proteins and to facilitate the design of proteins to bind to particular sequences (see Box 4-1). The structures of many different protein–DNA complexes have been solved by X-ray crystallography, including several examples of the same protein bound to different DNA sequences, as shown in Figure 4-2.

Figure 4-2A shows binding interactions of the native zif268 with its preferred binding sequence. Note that the protein's three fingers are numbered in reverse of the numbering of the positions in the binding site. Finger 3 (stippled) interacts with base-pairs 1–4, finger 2 (gray) with base-pairs 4–7, and finger 1 (blank) with base-pairs 7–10. In addition, note that each finger interacts with four base pairs but the spacing between fingers is only three base pairs so that there is one overlapping position between each adjacent pair of fingers: Position 4 interacts with both fingers 2 and 3 and position 7 interacts with both fingers 1 and 2. Finally, note that the positions in each finger are numbered relative to the structural alignment of that finger with the whole family of zinc finger proteins. In the canonical interaction, position −1, which is just before the α helix, interacts with base-pair 3 of each binding subsite (positions 3, 6, and 9 in the entire binding site), position 2 interacts with base-pair 4 of the subsite (positions 4, 7, and 10), position 3 interacts with base-pair 2 (positions 2, 5, and 8 of the whole site), and position 6 interacts with base-pair 1 of each subsite (positions 1, 4, and 7). In the examples shown in Figure 4-2, fingers 2 and 3 and their interacting base pairs are constant. Four different variants of finger 1 are shown in complex with two (or three in the case of protein RADR) different DNA sequences. In most cases, the affinity for the two sequences is quite similar; it is less than 10-fold different, except for DSNR-GCGT, which has a much lower affinity than its affinity to GACC, and RADR-GACC, which has a much lower affinity than its affinity to the other two sites (GCGT and GCAC). Wild-type zif268 has very similar affinity to sites with either T or G in position 10; thus, both GCGG and GCGT are considered consensus sites.

For wild-type finger 1 (RDER), the change in binding site from GCGT to GCAC results in the loss of several contacts with the bases. Glu2 is no longer seen to make a van der Waals contact with position 8, Asp2 does not contact position 10, and Arg-1 makes only one hydrogen bond to the A at position 9. These losses are partially compensated for by additional contacts with the backbone and with additional hydrogen bonds between the amino acids so that the decrease in affinity is not as large as would be expected from loss of base contacts alone.

In wild-type zif268, Asp2, besides hydrogen bonding to A (or C) at position 10, makes hydrogen bonds to Arg-1, helping to stabilize its contacts with G9. The D20A

Box 4-1. Recognition Code for Zinc Finger Proteins

Zinc finger TFs are highly modular, each containing multiple fingers and each finger specifically interacting with three or four base pairs of DNA. The order of the fingers can be switched and the binding-site sequence can change in a coordinated manner. Therefore, it seemed plausible that if the specificities of individual fingers could be determined, sets of fingers could be linked together in a specific order to design a protein with a desired specificity. Several groups have worked independently to determine the specificities of individual fingers using a variety of techniques. Zinc fingers are very abundant in higher eukaryotes; in fact, they are the most abundant family of TFs in mammalian and other genomes. Thus, determining the specificities of naturally occurring zinc finger TFs gives a large repertoire of binding specificities. This can be easily accomplished using SELEX (systematic evolution of ligands by exponential enrichment), a procedure that selects preferred binding sites from random libraries (see Chapter 5). But variant zinc finger proteins that do not occur in nature can also be generated and their specificities can be determined in the same manner. This may not identify proteins that prefer all possible binding sites, but a complementary method called phage display can select for proteins with high affinity for any particular DNA site (see Box 8-2). In this procedure, the protein is randomized (not the entire protein—just the amino acids that interact directly with the DNA and determine its specificity) and fused in a phage genome to the gene for a coat (or outer) protein of the phage. That protein will then be displayed on the surface of the phage where it can interact with DNA, and the DNA sequence that encodes the protein is internal to the phage. This links the protein sequence of the zinc finger protein to its DNA sequence and allows for selection and amplification of proteins with desired characteristics. An initial phage population containing the randomized zinc finger protein can be selected for those that bind to specific DNA sequences and then reinfected into bacteria, where they generate a new phage population that is now enriched for proteins with desired sequence-binding preferences. Typically after a few rounds of selective enrichment, the resulting phage are sequenced to determine the amino acids in the zinc finger that confer the desired specificity.

The information about the binding-site preferences of many different protein sequences can be combined into a recognition code for zinc finger proteins. Box Figure 4-1 shows such a recognition code from a review by Wolfe et al. (2000). This code assumes the canonical interaction geometry first observed in EGR1 (zif268) (see Figs. 3-6A and 4-2A), where the amino acids at positions 6, 3, and −1 (with respect to the α helix) interact with positions 1, 2, and 3 of the binding site, respectively. Position 2 of the protein can also interact with position 4 of the site, which overlaps with the site for the adjacent finger (unless there is none). The code represented by the table shows the preferred base, listed across the top, for each position in the site according to the amino acid at the corresponding position in the protein. For example, a G in the first site position is most often observed with proteins having an Arg at protein position 6, although Lys is also observed for that interaction. The asterisks indicate that both interactions have been observed in crystal structures, and the fact that Arg is in bold indicates it is the most common interaction. Question marks indicate uncertainty in the interaction.

Several aspects of the code are worth noting. One is degeneracy; for example, both Arg and Lys at protein position 6 can interact with G at site position 3, or His at protein position 3

Box 4-I. (*Continued.*)

Box Figure 4-I. Recognition code for zinc finger proteins (Wolfe et al. 2000).

interacts with both A and G at site position 2. Another aspect is that the interaction preferences are position specific. For example, although interactions with G at site positions 1 and 3 are both predominantly with Arg (at the corresponding protein position), at site position 2 a G interacts predominantly with His, and in fact, Arg is almost never observed in that interaction. This reflects the differences in geometry of the protein–DNA interface at those different positions. The code is valuable and has been used to design zinc finger proteins with desired specificities, but it is also limited in its utility. For example, the assumption of the canonical interaction between protein and DNA is not always correct (see Fig. 4-2). Furthermore, the code is inherently qualitative—it does not try to predict differences in affinities for different sequences. Generally, TFs bind to several different sequences with high, but variable, affinities, and knowing how the affinity changes with sequence is important for modeling regulatory interactions and designing proteins with desired properties. Quantitative specificity is a major focus of the remaining chapters of this book; Chapter 8 describes work to develop quantitative recognition models for protein–DNA interactions.

variant (Fig. 4-2B), which replaces Asp2 with Ala2, loses both of those contacts. Instead, Glu3 hydrogen bonds to Arg-1 to stabilize it and also makes a hydrogen bond with C8, replacing a previous van der Waals contact. In addition, Arg-1 now makes an additional hydrogen bond with G10. The result is that the D20A variant binds to the wild-type site with nearly the same affinity as the wild-type protein. Although the wild-type protein does not bind as well to GCTG (structure not known) than it does to GCGG, D20A binds nearly as well to both sites. Figure 4-2B shows that although the two hydrogen bonds with G9 are lost, a water molecule is fixed that forms hydrogen bonds among Arg-1, Glu3, and A9.

   Protein variant DSNR was selected to bind to the sequence GACC and makes hydrogen bonds to each of the four positions (Fig. 4-2C), similar to wild-type protein bonds to the wild-type binding site. Asn3 makes two hydrogen bonds to A8, Asp-1 makes a hydrogen bond to C9, and Ser2 makes a hydrogen bond to G10. In addition, some water-mediated contacts are not shown in the figure. If all of those bases are changed to the wild-type site (ACC to CGT), DSNR binding affinity is reduced. Asn3 now only makes a single hydrogen bond to C8, Asp-1 does not make any contacts with the DNA, and Ser2 makes water-mediated hydrogen bonds to A10 and an additional contact with the backbone.

Protein variant RADR was selected to bind the sequence GCAC, which it binds with nearly the same affinity as its affinity to the wild-type sequence GCGT (Fig. 4-2D). The wild-type sequence is bound very similarly as by the wild-type protein, except that Asp3 does not contact the bases and only hydrogen bonds to Arg-1; Ala2 makes a van der Waals contact with the T at position 11, extending the binding site by an additional base. Binding to the GCAC site loses the two hydrogen bonds between Arg-1 and G9 but gains a water-mediated bond between Asp3 and A8. It binds with much-reduced affinity to the site GACC, which only maintains the Arg6 contact with G7; all of the other direct base contacts are lost.

Ndt80

Ndt80, a yeast protein, is a member of the c-Rel homology family of TFs (see Chapter 3 and Fig. 3-8A). It regulates many key genes in sporulation and has been shown to bind to the middle sporulation element (MSE) whose consensus sequence is GNCRCAAAW (N = any base; R = A or G; W = A or T) (see Table 7-1). The structure of the complex has been solved for binding to consensus MSE site GACACAAAA (Fig. 4-3) and to 10 different variants of that site. In most cases, the change in binding affinity of a specific variant is between twofold and 10-fold, although there are a few with much larger effects. For the examples here, only the consensus structure is shown; the changes observed for the different variants are described in the following paragraphs.

If G1 is changed to either A or C, the interaction with Arg136 becomes lost and it rotates up to make contacts with the backbone. The A change has nearly the same affinity as the consensus site, whereas the C change reduces affinity by approximately threefold. This is less than one might expect for the loss of both hydrogen bonds to G, but the new backbone contacts partially compensate. There is also a distortion in the backbone between G and A in the consensus structure, which is relieved in the mutant. If A4 is changed to G, which is consistent with consensus R at that position, affinity decreases by approximately threefold. If it is changed to T, affinity decreases by more than fivefold. In both cases, there is loss of van der Waals contact between Arg111 and T at position 4. Approximately the same magnitude effect is observed if A6 is replaced by T. If C5 is changed to G, there is an ~100-fold reduction in affinity, but the structure of that complex has not been solved. There is a solved structure

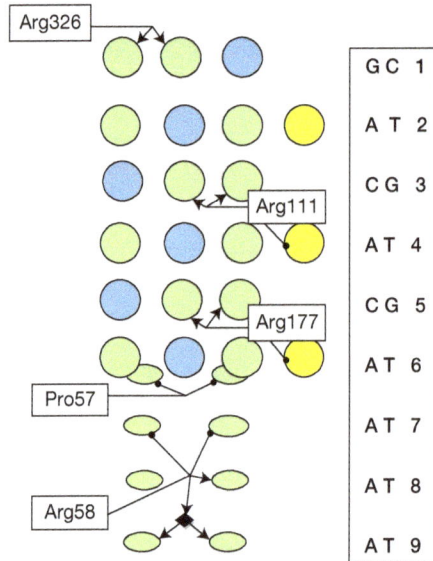

Figure 4-3. Interface between Ndt80 and an MSE binding site (Lamoureux and Glover 2006). (Small green ovals) Hydrogen-bond acceptors in the minor groove. For position 6, both major and minor groove contacts are shown; for positions 7–9, only minor groove contacts occur and thus the major groove is not shown. All other symbols are as used in previous figures.

for replacing C5 with T, but there are conflicting reports about its effect on binding affinity; in one report, it has an approximate fourfold effect and in another an approximate 50-fold effect. The effect on the structure is rather large, with Arg177 rotating by ~180°, where it displaces some well-ordered water molecules in the consensus structure.

The AT base pair at position 6 is contacted in both the major groove, by a van der Waals contact to Arg 177, and in the minor groove by additional van der Waals contacts to Pro57. Arg58 has a complicated set of interactions with all positions 7–9 in the minor groove: It makes van der Waals contacts with the AT pair at position 6, a direct hydrogen bond to T at position 7, and a water-mediated hydrogen bond to the AT pair at position 9. Replacing any of the AT pairs with TA has an approximate threefold effect on affinity, but it is not clear why this occurs because very similar contacts should remain. Replacing any of the AT pairs with GC pairs has a larger effect on binding, because the additional hydrogen bond donor in the minor groove of GC base pairs (see Fig. 2-2C) disrupts the consensus interactions as well as four well-ordered water molecules observed in the minor groove. Positions 7–9 also have a somewhat narrower minor groove than standard B-form DNA, which is typical for poly(A) tracks, and that may also contribute to binding affinity.

The Lac Repressor

The Lac repressor, the first regulatory system discovered, led to general models of TFs regulating gene expression (Chapter 1). The Lac repressor has been studied biochemically and genetically for many years, but structural information has been difficult to obtain. Using nuclear magnetic resonance (NMR) techniques, Romanuka et al. (2009) have determined the interactions at the protein–DNA interface for the three natural operator sequences: O1, O2, and O3. O1 is the primary site of the Lac repressor binding to control expression of the *lac* operon (see Fig. 1-1), and O2 and O3 are upstream and downstream, respectively (relative to the direction of the *lac* operon), and were discovered years later. They contribute to regulation because the tetrameric Lac repressor can bind to O1 and one of the other operators simultaneously, creating a loop of the intervening DNA that contributes to stability of the protein–DNA interaction. The binding site is 17 bp with a palindromic consensus sequence TTGTGAGCSGCTCACAA (*S*, the middle base, corresponds to G or C; the left and right halves are complementary) that binds the dimeric repressor protein. Figure 4-4 shows the overall interaction of the Lac repressor DNA-binding domain with DNA. Besides the HTH motif that binds in the major groove and provides many direct base contacts, another "hinge helix" binds in the minor groove at the center of the palindromic site (see Fig. 3-4B). The minor groove is widened and the DNA is bent to accommodate the hinge helices from both monomers. Leu56 from each monomer intercalates into the DNA helix on both sides of the center GC base pair, contributing to bending of the DNA helix.

The left half of O1 is a perfect match to the consensus sequence, but all of the other half-sites contain at least one mismatch (the "left half" of each site can be defined by the strand that contains G in the middle position). Figure 4-4A shows the interactions of the Lac repressor with O1L (left half of O1), not including the intercalated Leu56 between position 8 and the center base pair. There are numerous contacts, at least one to each base pair, including both hydrogen bonds and van der Waals interactions. Arg22 provides hydrogen bonds to two adjacent bases, T2 and G3, whereas His22 and Ser16 provide van der Waals contact with T4. G3 and T4 occur in every half-site except O3R (the right half of operator O3) (see below) and the contacts with Arg22 and Ser16 are maintained. Those contacts and additional contacts with operator positions 5–8 can explain the reason that the protein has a large maximum specificity of $\sim 10^{6}$-fold between consensus sequence and nonspecific binding. At the same time, the protein binds to sequences that differ from the consensus at one to a few positions, with relatively modest effects on binding. The structures obtained for the different natural half-sites—O1R, O2L, O2R, and O3L—show how the protein adapts to altered DNA sequences to maintain relative high affinity, even though some important contacts observed in the consensus sequence are rearranged or eliminated.

Figure 4-4B shows the sequence of O1R (shown in the same orientation as O1L, with the center of the operator at position 9), which has two changes from the

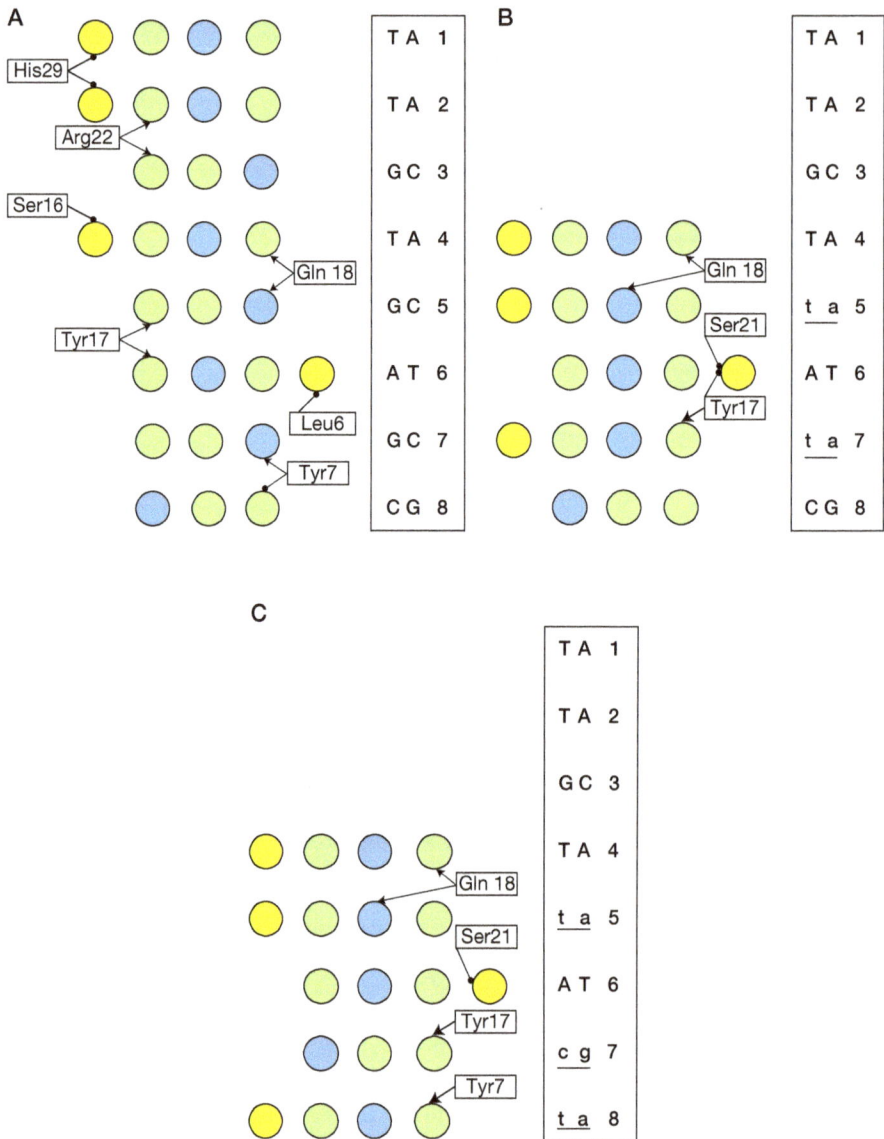

Figure 4-4. The Lac repressor bound to several different binding sites (Romanuka et al. 2009). (A) Lac repressor interactions with the left half of operator 1 (O1). Not shown are the backbone contacts or the central base and Leu56, which intercalates between position 8 and the central GC base pair. (B) The Lac repressor with the right half of O1. Differences in the DNA sequence are shown in lowercase and underlined. Only those changes in interactions with the protein are shown, but in addition, the previous interactions with Tyr7 and Leu6 are lost on this half-site. (C) Interaction of the Lac repressor with the right half of O2. Differences in the site sequence from the left half of O1 are in lowercase and underlined. Only those changes from the interactions seen in the left half of O1 are shown.

consensus (lowercased and underlined). Positions 1–3 are the same as those for O1L and have the same interactions with the protein, which are not shown. The changes at positions 5 and 7 cause several rearrangements of the contacts between protein and base pairs. Gln18 still makes two contacts with positions 4 and 5, but they are shifted compared with the consensus sequence. Tyr17 loses both of its previous contacts, but makes two entirely new contacts, and Ser21 adds a new van der Waals contact to position 6. The previous interactions with Leu6 and Tyr7 are both lost and there are now no direct contacts to position 8.

O2L is the same as O1L except that position 1 is changed from T to A. This causes the loss of both van der Waals contacts with His29, but all of the other contacts remain the same. Figure 4-4C shows the contacts with O2R, which has three changes from the consensus. The sequence of positions 1–3 and their contacts with the protein are the same as those for the consensus and are not shown. The TTA sequence for positions 4–6 is the same as that for O1R and the same contacts with Gln18 and Ser21 are seen. But now, Tyr17 makes contact only to G at position 7 and Tyr7 makes a new hydrogen bond to A at position 8.

O3L is the same as the consensus except that positions 1 and 2 are changed from TT to CA. The contacts of those positions with His29 and one of the hydrogen bonds to Arg22 are lost, but otherwise the structure is essentially the same as that for O1L. O3R has four changes from the consensus sequence and nearly all direct base contacts are lost, consistent with O3 binding with much-reduced affinity to the Lac repressor, ~1000-fold lower than that to O1.

STRUCTURES OF NONSPECIFIC BINDING

Binding of the Lac repressor to O3R provides a view of how we expect nonspecific binding to occur. Protein contacts are made almost exclusively to the backbone, with many of them the same contacts that are made in sequence-specific binding, but there are also some new backbone contacts, including those to amino acids that bind directly to base pairs in other structures (von Hippel 2004; Romanuka et al. 2009). The helices providing that fit in the major and minor grooves are rotated away so that normal contacts cannot be made. A very similar story is seen with nonspecific binding of the Matα2 homeodomain protein (Aishima and Wolberger 2003). In a crystal structure that contained two consensus binding sites and two proteins bound there, there were also two additional proteins bound to DNA in a nonspecific mode. As with the Lac repressor bound to O3R, the recognition helix was rotated away from the DNA so that most of the normal contacts could not be made.

In these two different nonspecific examples, there are a couple of direct contacts with base pairs, some using alternative amino acids from those that are normally used for specific binding, but the majority of the contacts are to the backbone. Because there are very few, if any, direct contacts are to base pairs in the nonspecific binding mode, it is easy to imagine how the protein can slide along the DNA until it comes to a sequence that provides several favorable contacts, where it can then rotate to make those bonds and maintain a stable interaction.

LESSONS ON SPECIFICITY OF TRANSCRIPTION FACTORS

The various structures of protein–DNA complexes provide insights into how specificity is obtained and modulated. The EcoRI restriction site has a very high degree of fine-scale specificity, where any change to the binding site reduces affinity enormously. This is accomplished with multiple contacts to each base pair plus the distortion to the DNA backbone and the rigidity of the protein itself, where numerous interamino acid contacts prevent it from easily adapting alternative interactions with different sequences. The TFs all have similar degrees of high maximum specificity—$\sim 10^6$-fold compared with nonspecific binding—but this specificity is spread out over a longer binding site and thus individual changes are usually tolerated with relatively minor reductions in affinity.

Different amino acid sequences prefer different binding-site sequences through the complementary arrangements of hydrogen-bond donors and acceptors as well as van der Waals contacts, water-mediated contacts, extensive backbone contacts, and interactions between the amino acid side chains. Although all of these interactions allow for large maximum specificities, the flexibility of the interface to alter the web of contacts also allows for many different sequences to have affinities not too different from the preferred binding site. Each protein accomplishes this task in its own way, making rearrangements that are allowed by its particular structure. If a single base pair in the site is changed, the rearrangements in the protein are only local. This explains why models for specificity that assume the positions contributing independently to binding work reasonably well (see Chapter 7) and at the same time explains why simple rules for recognition have not been found.

REFERENCES

Aishima J, Wolberger C. 2003. Insights into nonspecific binding of homeodomains from a structure of MATα2 bound to DNA. *Proteins* **51:** 544–551.

Elrod-Erickson M, Benson TE, Pabo CO. 1998. High-resolution structures of variant Zif268-DNA complexes: Implications for understanding zinc finger–DNA recognition. *Structure* **6:** 451–464.

Garvie CW, Wolberger C. 2001. Recognition of specific DNA sequences. *Mol Cell* **8:** 937–946.

Lamoureux JS, Glover JN. 2006. Principles of protein–DNA recognition revealed in the structural analysis of Ndt80-MSE DNA complexes. *Structure* **14:** 555–565.

Luscombe NM, Laskowski RA, Thornton JM. 2001. Amino acid-base interactions: A three-dimensional analysis of protein–DNA interactions at an atomic level. *Nucleic Acids Res* **29:** 2860–2874.

Miller JC, Pabo CO. 2001. Rearrangement of side-chains in a Zif268 mutant highlights the complexities of zinc finger-DNA recognition. *J Mol Biol* **313:** 309–315.

Pingoud A, Jeltsch A. 1997. Recognition and cleavage of DNA by type-II restriction endonucleases. *Eur J Biochem* **246:** 1–22.

Rohs R, West SM, Sosinsky A, Liu P, Mann RS, Honig B. 2009. The role of DNA shape in protein-DNA recognition. *Nature* **461:** 1248–1253.

Romanuka J, Folkers GE, Biris N, Tishchenko E, Wienk H, Bonvin AM, Kaptein R, Boelens R. 2009. Specificity and affinity of Lac repressor for the auxiliary operators O2 and O3 are explained by the structures of their protein-DNA complexes. *J Mol Biol* **390:** 478–489.

von Hippel PH. 2004. Completing the view of transcriptional regulation. *Science* **305:** 350–352.

Wolfe SA, Nekludova L, Pabo CO. 2000. DNA recognition by Cys_2His_2 zinc finger proteins. *Annu Rev Biophys Biomolec Struct* **29:** 183–212.

CHAPTER 5

Binding Affinity, Cooperativity, and Specificity

C HAPTERS 2–4 FOCUSED ON STRUCTURAL ANALYSES OF DNA, protein, and their complexes. General properties as well as many common variations of DNA structure were described in Chapter 2. Overall structural features of proteins were described in Chapter 3, with an emphasis on transcription factor (TF) families and the diverse and distinctive ways in which they bind to DNA. Chapter 4 focused on structural analyses of specificity—how amino acids on the surface of proteins can interact with DNA to distinguish between different DNA sequences, binding to some with much higher affinity than to others. That chapter further showed the plasticity at the protein–DNA interface, where contacts can be rearranged so that different sequences can be bound with similar affinities through alternative interaction patterns.

Chapter 5 begins an analysis of the thermodynamics of protein–DNA interactions. The emphasis is on experimental methods for measuring binding affinity of a protein for a particular DNA sequence, possible effects of other proteins binding nearby, and the distribution of binding affinities for the entire set of possible binding-site sequences. Measurements of protein–DNA-binding affinities provide a complementary approach to structural studies, determining the quantitative effects of alternative interactions that can only be qualitatively understood from the visualization provided by the structures. Furthermore, experimental measurements of binding affinities can be determined using high-throughput (HT) techniques that allow for large-scale comparisons and development of quantitative models of specificity.

BINDING AFFINITY

At its most basic level, the binding of a TF to a site on double-stranded DNA (dsDNA) is a bimolecular interaction that can be described with two parameters, an on rate,

k_{on}, and an off rate, k_{off} (Fig. 5-1). The rate of complex formation is determined by (1) the concentration of the two components, which determines how often they encounter one another, and (2) the probability of the components forming a complex when they meet. If the two components were simply floating independently in the cell, the frequency of their encounters would be limited by diffusion—how rapidly they explore the volume of the cell. But because all proteins can bind to DNA non-specifically and slide along the DNA searching for good binding sites, they can encounter specific sites faster than would be allowed by simple three-dimensional diffusion, as is described in more detail in Chapter 6. The rate at which the complex forms inside a cell can also be reduced by competition with other proteins, which sometimes make specific sites inaccessible to the TF and completely inhibit their interaction.

A

$$TF + S \underset{k_{off}}{\overset{k_{on}}{\rightleftharpoons}} TF \cdot S$$

B

$$[TF][S]k_{on} = [TF \cdot S]k_{off}$$

$$K_D = \frac{K_{off}}{K_{on}} = \frac{[TF][S]}{[TF \cdot S]} = \frac{1}{K_A} = e^{\Delta G°/RT}$$

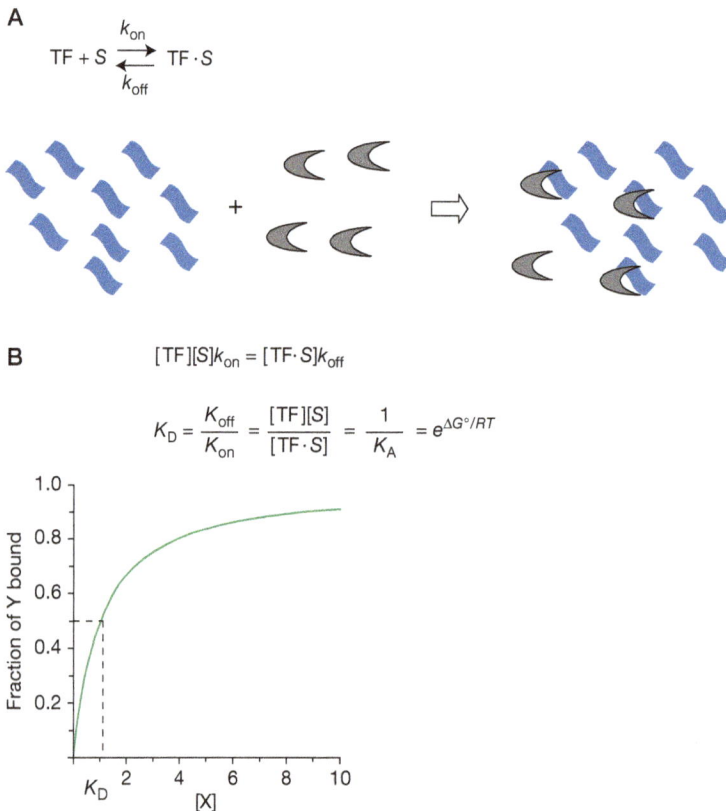

Figure 5-1. (A) Reaction equation and diagram of interacting TF (gray crescents) and DNA (blue lines). (Adapted from Stormo and Zhao 2010). (B) At equilibrium, the rates of complex formation and dissolution are equal. The dissociation constant K_D is equal to the free concentration of one component [X] (either DNA or protein), at which the other component Y is half bound.

Off rate is determined by the binding energy of the complex. As described in Chapter 4, complexes between protein and DNA involve a variety of noncovalent bonds, including electrostatic interactions, hydrogen bonds, and van der Waals contacts. At any given instant, there is a probability that thermal motions will be sufficient for the two components to separate and lose contact, characterized by an off rate. Because the complex dissociates randomly with a specific rate, it can also be characterized as having a half-life, or the time at which half of the complexes will have dissociated: half-life $= \ln2/k_{off}$ ($= 0.69/k_{off}$).

The ratio of off rate to on rate is referred to as the dissociation constant K_D and is a useful measure of the affinity between protein and DNA. Its reciprocal, the association constant K_A, is also frequently used to describe affinity of a protein–DNA interaction. Both constants are related to the "Gibbs standard free energy of binding, $\Delta G°''$ (Fig. 5-1B), which is described in more detail in Chapter 6.

The fraction of DNA that is bound, or equivalently, the probability that any specific DNA sequence S will be bound at a particular time, is

$$\text{Pr}(S \text{ bound}) = \frac{[TF \cdot S]}{[TF \cdot S] + [S]} = \frac{K_A[TF]}{K_A[TF] + 1} = \frac{1}{1 + K_D/[TF]}, \qquad (5\text{-}1)$$

where Pr refers to the probability, TF is the transcription factor, and the brackets indicate concentrations. It is important to recognize that [TF] refers to the concentration of the free (that is, unbound) TF, not the total concentration of the TF, some of which will be in the complex with DNA.

The last form of Equation 5-1, using K_D, shows that the DNA will be half bound (Pr $= 0.5$) when the concentration of free TF, [TF], is equal to K_D. But for many purposes, the form of the equation using K_A is more useful, as shown below in the description of cooperativity.

There are a variety of experimental methods for determining K_D for a specific protein and DNA sequence, and the popularity of different methods has changed over time as technological advances have allowed more convenient and accurate methods. Most methods determine the equilibrium concentrations of the complex and one of the two free components as the concentration of the other component is varied, which gives a plot as shown in Figure 5-1B. For example, if the protein concentration greatly exceeds that of DNA, its free concentration ([X] in the figure) will be very nearly equal to its total concentration. Then, the fraction of DNA in the DNA–protein complex can be measured (plotted on the vertical axis of the figure), and the concentration of protein at which DNA is half bound is K_D. Of course, the experiment could also be performed with DNA in excess and the fraction of bound protein measured, but labeling of DNA, either radioactively or fluorescently, is generally easier and leads to more sensitive and accurate measurements.

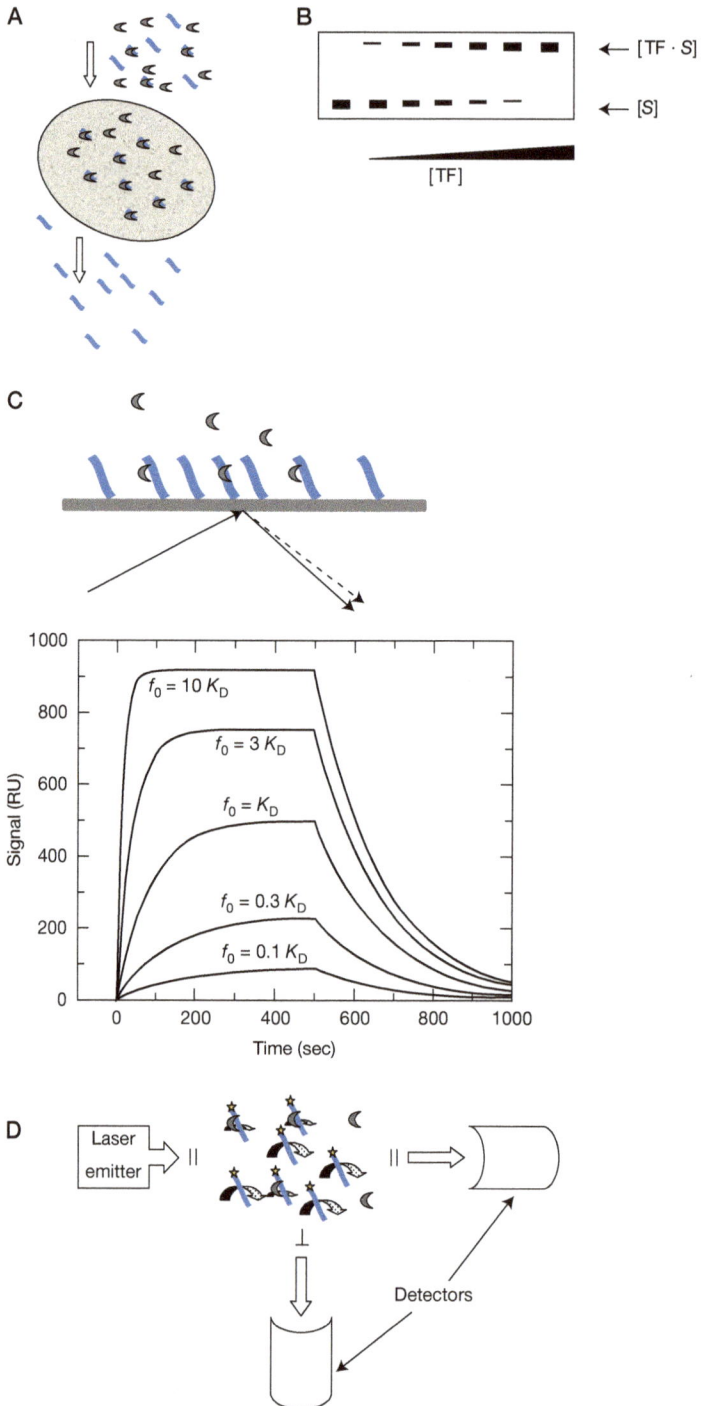

A

B

←— [TF · S]

←— [S]

[TF]

C

$f_0 = 10\ K_D$

$f_0 = 3\ K_D$

$f_0 = K_D$

$f_0 = 0.3\ K_D$

$f_0 = 0.1\ K_D$

Signal (RU)

1000

800

600

400

200

0

0   200   400   600   800   1000

Time (sec)

D

Laser
emitter

Detectors

Figure 5-2. (*See facing page for legend.*)

Experimental Methods to Determine K_D

Filter-Binding Assays

The earliest experiments were filter-binding assays (Fig. 5-2A), which rely on the fact that proteins stick to nitrocellulose filters, whereas DNA (and this can also be used for RNA) flows through the filter unless bound by a protein. DNA can be easily labeled to high specific activity with radioactive phosphorous (^{32}P), allowing easy determination of the fraction of total DNA bound to the filter at different protein concentrations.

Electrophoretic Mobility-Shift Assays

The next generation of methods was electrophoretic mobility-shift assays (EMSAs), also referred to as gel- or band-shift assays (Fig. 5-2B). As with the filter-binding

Figure 5-2. Techniques for determining binding affinity. (A) Filter-binding experiment. DNA (blue lines) and protein (crescents) are mixed to form a complex. The mixture is flowed through (arrows' direction) a filter (stippled oval). Protein sticks to the filter along with any DNA that is bound to the protein. Unbound DNA flows through the filter. The fraction of DNA bound to the filter at different protein concentrations can be used to measure K_D of the interaction. The experiment is typically performed with a large excess of protein (DNA concentration is well below K_D, whereas protein concentration covers a range from below to above K_D), so that the amount of free protein [TF] is approximately equal to the amount of total protein. (B) Electromobility shift assay (EMSA). In each lane of the gel (outer rectangles), a mixture of DNA and protein is added. When voltage is applied, DNA moves down the gel, but it moves more slowly if it is bound by protein (upper bands) than if it is unbound (lower bands). By using increasing concentrations of protein (from left to right), one can determine the fraction of DNA bound at different concentrations of protein. The experiment is usually performed at protein excess so that the amount of free protein is approximately equal to the total amount of protein. In this figure, the left lane has no protein and the rightmost lane has enough protein that DNA is essentially all bound. K_D would be protein concentration in the middle lane, where DNA is 50% bound. (C) Surface plasmon resonance (SPR) experiment. Here, DNA is attached to the surface. Light shining on the surface is reflected at a certain angle (solid arrow). When a protein binds to the DNA, the angle of reflection is changed (dashed arrow). The amount of change depends on the amount of protein bound. Protein is added at a certain concentration, and absorption that occurs over time can be observed and the on rate determined from that (lower figure). The total reflectance change (vertical axis: Ru [relative units]) will eventually plateau when the system has reached equilibrium; the height of the plateau varies with protein concentration (here, shown relative to the K_D of binding). The solution containing the protein can then be removed and replaced with pure buffer, allowing the measurement of the off rate as proteins dissociate from DNA (lower figure). (D) Fluorescence anisotropy experiment. DNA is labeled with a fluorescent dye (stars), which is excited by laser light that has been polarized (parallel line filter). The dye emits light that is also polarized and is detected by two detectors, one with the same polarization as the excitation light (right detector) and another perpendicular to the polarization of the excitation light (lower detector). The DNA will rotate in solution (indicated with stippled arrows), and if emission occurs before it has rotated very much, this will be detected by the parallel detector. If emission occurs after the DNA has rotated sufficiently, this will be detected by the perpendicular detector. Protein binding to DNA decreases the rate of rotation (smaller arrows) such that one can determine the fraction bound by comparing amount of light detected by the parallel and perpendicular detectors as a function of protein concentration.

assays, protein is in excess and DNA is labeled either radioactively or with fluorescent dyes. This assay relies on the fact that DNA will migrate through a gel when voltage is applied and move toward the positive end due to its high negative charge. The rate of migration is decreased when the DNA is bound to protein; thus, after the reaction has come to equilibrium, the reaction mixture is placed at the top of the gel and voltage applied. Once the complex enters the gel, it essentially becomes completely stable. Complexes will migrate through the gel for many times the half-life of the complex in solution, meaning that the ratio of free DNA to the DNA–protein complex, which can be measured in the gel, is essentially a "snapshot" of the equilibrium condition. By varying the amount of protein added, one can determine K_D from a plot, as in Figure 5-1B. A variation of this idea is to add unlabeled ("cold") competitor DNA to the reaction at varying concentrations. By measuring how much of the bound fraction of labeled DNA decreases as a function of the amount of unlabeled DNA, K_D relative to that of the competitor can be determined. This is especially useful if the unlabeled DNA has a known K_D and one wants to measure how well it competes with the labeled DNA, which contains a different sequence with an unknown K_D.

Sometimes, the kinetic rate constants k_{on} and k_{off}, as well as K_D, are of interest. k_{off} is easiest to measure. After the reaction has reached equilibrium, the solution is diluted so that no new complexes will form, and samples are extracted at various time points to measure how much the fraction bound has decreased over time, giving a half-life. This could be done by either of the methods described above, where the fraction bound is determined at different time points, but new technologies allow determination of both k_{on} and k_{off} rates directly.

Surface Plasmon Resonance

Surface plasmon resonance (SPR) relies on the fact that the angle of reflectance from a thin metal surface, usually gold, changes when something is bound to the other side (Fig. 5-2C, top), and the amount of change depends on the molecular weight of what is bound. Originally used in biology to study protein–protein or protein–ligand interactions, this method has also been adapted for use in protein–DNA interactions. Typically, dsDNA (although it could also be ssDNA or RNA) is attached directly to the surface of the gold and immersed in a binding buffer solution. Added protein binds the DNA sequences on the surface, altering the reflectance angle. The change in reflectance can be monitored through time until the solution reaches equilibrium, at which point the angle remains constant; an on rate is determined by knowing the concentration of DNA and protein in the solution and the time required to reach equilibrium (Fig. 5-2C, bottom). The buffer can then be replaced without protein so that as the protein dissociates from the DNA, it is diluted and unlikely to rebind. Again, the time course can be followed until the reflectance returns to its original level (before protein binding), from which the off rate can be determined (Fig. 5-2C, bottom).

The ratio of on and off rates gives K_D. This method provides the additional information of those rates directly. SPR can be applied in parallel to measure K_Ds of many different DNA sequences simultaneously by having multiple different spots on the surface, each with its own DNA attached, and the change in reflectance measured for each spot independently by scanning across the array with the laser (discussed in more detail below).

Fluorescence Anisotropy

Changes in fluorescence anisotropy can also be used to determine the K_D of the protein–DNA interactions. This method is based on the fact that a fluorescent dye, when stimulated with polarized light, will emit light with the same polarization relative to the orientation of the dye molecule. As the dye rotates in solution, its orientation changes and so does its polarization. By measuring the emission polarization, both in the plane parallel to the original polarization and in the plane perpendicular to it, one can determine the rate of rotation of the dye, relative to its emission half-life. One must choose dyes whose half-life of the excited state, before emission, is longer than the typical rotation velocity of the molecule, and there are many such dyes to use. For the study of DNA–protein interactions, dye is attached to DNA in such a way that its orientation depends on the orientation of the DNA itself (it is attached so that it does not rotate independently of the DNA). The time for complete depolarization, when the output polarization is random compared to the input, depends on how fast the DNA rotates in solution. When protein is bound to DNA, it rotates more slowly because it is bulkier (Fig. 5-2D). By varying protein concentration and measuring the amount of depolarization over time, the fraction of bound DNA can be derived. The main advantage of this method over such methods as filter binding or EMSA, which also determine the fraction of DNA bound as a function of protein concentration, is that this experiment is performed in solution under equilibrium conditions, avoiding artifacts that may occur with nonequilibrium assays.

BINDING COOPERATIVITY

Often, there are binding sites for multiple TFs in close proximity to one another in a segment of DNA (see the figures in Chapter 1). If the factors bind independently, K_Ds for each protein binding to its own sites can be used to model the behavior of the entire system. Any deviation from independent binding is referred to as cooperativity, although it is common to separate positive from negative cooperativity. In positive cooperativity (sometimes just referred to as cooperativity), binding of one factor increases binding of the other compared to what it would be in its absence. In negative cooperativity (sometimes called anticooperativity), binding of one factor decreases binding of the other factor; in the extreme case, binding of two factors might be mutually exclusive in that only one can bind at a time, although binding sites for both exist. Figure 5-3 shows an example of two factors, TF_1 and TF_2, with binding sites for each, S_1 and S_2, in close proximity. The reaction diagram shows that either factor might bind alone or they could both bind to form a complex with two proteins bound to the DNA. If K_1 and K_2 are the association constants for each factor binding to its site alone, and $K_{1|2}$ and $K_{2|1}$ are the association constants when the other factor is already bound, the association constant for the binding of both factors to DNA is

$$K_{1,2} = K_1 K_{2|1} = K_2 K_{1|2} = \omega K_1 K_2, \qquad (5\text{-}2)$$

where ω is the cooperativity constant. The total association constant must be independent of the order in which the factors bind. If $\omega = 1$, the two factors bind independently, if $\omega > 1$, they show positive cooperativity, and if $\omega < 1$, they show negative cooperativity. If their binding is mutually exclusive, then $\omega = 0$.

Equation 5-1 shows the probability of a DNA segment being bound, and obviously the probability of it not being bound is just $\Pr(S\ \text{unbound}) = 1 - \Pr(S\ \text{bound})$ because those are the only two states possible, bound and unbound. DNA segments that have multiple binding sites can exist in many more states. As shown in Figure 5-3, DNA with two binding sites can be in four possible states. DNA with three binding sites could exist in eight possible states. In general, DNA with n sites can exist in 2^n different states (although some sites may be mutually exclusive, reducing the number of possible states). The same approach can be used to calculate the probability of each state. Considering the example of Figure 5-3, we can label the different states

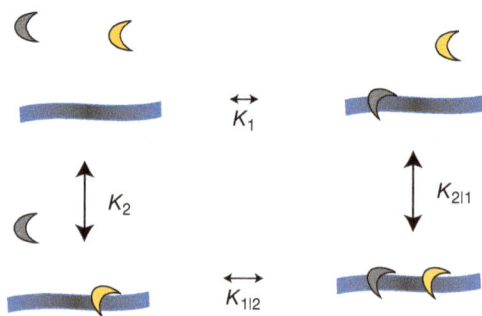

Figure 5-3. Cooperativity is the nonindependent binding of two proteins to DNA. Here, DNA has two binding sites, one for each protein (two colors). Proteins can bind in either order (top row or left column) and then both can be bound. Binding constants K are indicated for protein 1 being bound (K_1) or protein 2 being bound (K_2) separately. Also indicated are the binding constants for protein 1 if protein 2 is already bound ($K_{1|2}$) and for protein 2 if protein 1 is already bound ($K_{2|1}$). If the proteins bind independently, $K_1 = K_{1|2}$ and $K_2 = K_{2|1}$. If not, there is cooperativity between them such that binding of both proteins to the DNA can be described by $K_{12} = \omega K_1 K_2$, where ω is the cooperativity constant.

j_1, j_2, $j_{1,2}$, and j_0, corresponding to the binding of TF_1 alone, TF_2 alone, both TFs, and neither TF, respectively. The probabilities of each state are

$$Pr(j_1) = \frac{K_1[TF_1]}{1 + K_1[TF_1] + K_2[TF_2] + \omega K_1 K_2[TF_1][TF_2]},$$

$$Pr(j_2) = \frac{K_2[TF_2]}{1 + K_1[TF_1] + K_2[TF_2] + \omega K_1 K_2[TF_1][TF_2]},$$

$$Pr(j_{1,2}) = \frac{\omega K_1 K_2[TF_1][TF_2]}{1 + K_1[TF_1] + K_2[TF_2] + \omega K_1 K_2[TF_1][TF_2]},$$

$$Pr(j_0) = \frac{1}{1 + K_1[TF_1] + K_2[TF_2] + \omega K_1 K_2[TF_1][TF_2]}.$$

(5-3)

Note that the sum of the probabilities over all possible states is 1, as required, because each DNA molecule must be in one of those states. The denominator, called the partition function, is the sum over all possible states. The same approach can be used for any number of sites and states, although as described above, the number of possible states grows exponentially with number of sites. When the number of states becomes too large to enumerate in the denominator, the symbol Z is often used to indicate that sum, so that, for example, one could write $Pr(j_1) = K_1[TF_1]/Z$, and that would be true regardless of the number of possible states.

The probability that a particular site is occupied is the sum of probabilities of all states with that site occupied. For example, $Pr(S_1$ occupied$) = Pr(j_1) + Pr(j_{1,2})$. Note also that the probability that DNA is bound by at least one TF is not the sum over the binding probabilities for each site being occupied because that double counts the state with both TFs bound. The example in Figure 5-3 has two different TFs binding to DNA but it could also be the same TF binding to both sites. In that case, the equations above would change to have only a single [TF], and $[TF_1][TF_2]$ would be replaced by $[TF]^2$. K_1 and K_2 could still be different if the two sites S_1 and S_2 are different, or they could be the same.

The physical basis of positive cooperativity is usually due to favorable interactions between proteins, so that when one protein is bound, the successive protein gets favorable energy from binding to the DNA plus the favorable energy of the protein–protein interaction, such as shown in the example in Figure 3-8B. It is also possible to get positive cooperativity without direct interaction of proteins. For example, one protein can distort the DNA around its binding site to increase the affinity of the other protein binding nearby; although possible, this is less likely because DNA distortions tend to be short range. In supercoiled DNA, the change in local twist can have longer-range effects and induce positive (or negative) cooperativity at a distance. It is also possible for there to be indirect cooperativity between two proteins, the result of them both competing for binding with another common factor. For example, two or more TFs that bind near to each other, but which do not interact and would not be seen as cooperative in an in vitro binding assay, could display cooperativity in vivo if binding of both factors requires displacement of a nucleosome (see Fig. 1-3). When the nucleosome is present, none of the factors can bind. But when it is absent, they can all bind; thus, the system is more likely to be in states with all or none of the sites bound than with only a fraction of the sites bound, giving apparent cooperativity. Negative cooperativity could be due to adjacent proteins repelling each other, but is more likely due to the overlap of their binding sites on DNA. If they overlap only partially, it may be possible for both to bind (with reduced affinity), but if they overlap substantially, they are likely to be mutually exclusive.

Methods to Assay Cooperativity

Using Affinity Assay Methods

Methods for assaying affinity can sometimes be adapted to measure cooperativity. Filter binding does not distinguish between one or multiple proteins being bound. However, if the binding of each protein independently is determined first, and then the fraction of DNA bound by at least one protein (which is sufficient to cause filter binding) is measured when both proteins are added, it is possible to measure cooperativity. EMSA may give two distinct bound bands, one corresponding to both proteins binding and the other corresponding to either protein being bound (if one TF is much larger than the other, it may even be possible to distinguish independent binding of the two TFs). Again by first determining K_Ds for the proteins independently and then for the mixture of both proteins, it is possible to determine cooperativity. It is conceivable that SPR and fluorescence anisotropy methods could be adapted to determine cooperativity by determining independent affinities and then combined affinity. Alternatively, one might add one protein in saturating amounts, so that all of the DNA is bound by that protein, and then add the second protein in varying amounts to determine its K_D in the presence of the other protein compared to its K_D alone.

Nuclease Protection

An alternative method that not only detects binding to DNA but also localizes where the binding occurs, thereby distinguishing between different binding sites, is a nuclease protection or "footprinting" assay (Fig. 5-4). After a protein–DNA interaction has reached equilibrium, an enzyme such as DNase I (this can also be done with chemicals that attack the DNA), which cuts DNA nonspecifically, is added for a brief time. It will cut the DNA, which is labeled on one end, so that when the DNA is then run through a gel, it will be separated into fragments of different lengths, defined by where the cut occurs. The protein protects the DNA from cutting at the binding site, leaving a

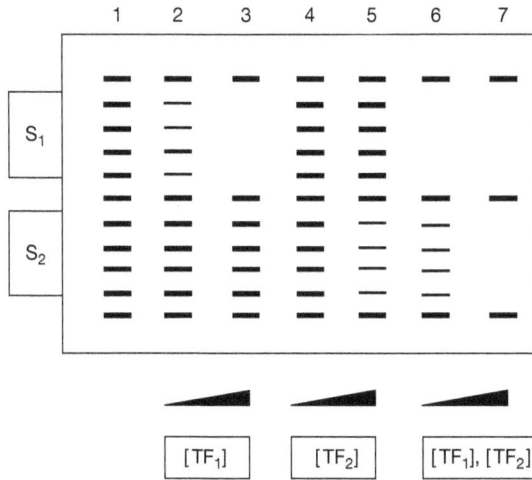

Figure 5-4. Footprinting method for determining binding affinity and cooperativity. Shown is the location of each binding site on the DNA. Lane 1 (leftmost) shows the cleavage pattern seen with no protein added for protection. Lanes 2 and 3 show the protection for increasing concentrations of TF_1, and lanes 4 and 5 show the protection for increasing concentrations of TF_2. Lanes 6 and 7 show the protection of both binding sites when the two TFs are added together in increasing concentrations. Cooperativity is observed because the two binding sites are protected to a higher degree at the lower concentrations than seen for each protein alone.

"footprint" where it is bound. By altering the concentration of the TF, the amount of protection afforded can be determined and a measurement of the K_D obtained.

The main advantage of nuclease protection over the methods described above for affinity assays is that each site on the DNA can be assayed independently. In Figure 5-4, each protein is added separately to determine individual K_Ds for each site. The combination of both proteins is then added and the protection of both sites observed. In this example, TF_1 binds with higher affinity than does TF_2, so that at low concentration, a partial footprint is seen, and at high concentration total protection is observed. Little protection is observed at low concentration for TF_2, but moderate protection is seen at high concentration. When both proteins are added together (at the same individual concentrations as in the previous lanes), there is apparent cooperativity because even at low concentration the TF_1 site is completely protected and the TF_2 site is partially protected. By careful quantification of the degree of protection over a range of TF concentrations, the K_Ds for each protein binding to its own site as well as the cooperativity factor ω can be determined.

Another advantage of footprinting assays is that they are easily used when both sites are bound by the same protein. Although Figure 5-4 shows lanes that specifically assay each protein individually, as was described above for using the other methods to determine cooperativity, this is not necessary. Because binding sites are assayed independently on the same DNA segment, K_Ds to each site and cooperativity can be obtained in a single titration experiment when the concentration of TF is varied over the important range.

BINDING SPECIFICITY

Specificity refers to the ability of some DNA-binding proteins to bind to different sequences with different affinities. Specificity of TFs is critical for their ability to regulate gene expression. Bacterial genomes typically contain millions of base pairs and many plant and animal genomes have billions of base pairs, all of which are potential binding sites. These numbers far exceed the number of TF proteins in the cell; thus, the factors must bind to a very small subset of potential binding sites. This requires that they have different affinities for different sequences and that the highest-affinity sequences are positioned within the genome where binding events can affect the expression of their regulated genes.

The specificity needed by a TF for a regulatory system to work can be estimated based on the size of the genome and the number of regulated genes. For example, the *Escherichia coli* genome has $\sim 5 \times 10^6$ base pairs (or $\sim 10^7$ potential binding sites when considering both strands of the DNA). If a TF were to regulate just one gene, it could do so if it were extremely specific for a 12-long sequence. Because $4^{12} > 10^7$, not all 12-long sequences can occur in the genome, and many of them will occur only once, whereas shorter sequences are likely to occur multiple times unless specifically selected against. If the TF were so specific that it had no affinity for any other sequences, a single copy would be sufficient. But real TFs have nonzero affinities for other sequences, including a nonspecific affinity for essentially any sequence that is typically 10^4- to 10^6-fold lower than to the preferred sequence. Using the example of a single regulatory site in *E. coli*, if the nonspecific affinity were 10^6-fold lower than to the regulatory site, a single protein would spend $\sim 90\%$ of its time bound nonspecifically and 10% bound to the regulatory site, because there are 10^7 nonspecific sites each with 10^6-fold lower affinity than the specific binding site. For the regulatory site to be occupied most of the time, there would need to be many more copies of the protein in the cell than the number of regulatory sites to overcome the "sponge" effect of nonspecific binding from the whole genome. In a genome such as the human genome, with more than 10^9 base pairs, the problem is much worse, and thousands of times more protein would need to be synthesized than the number of regulatory sites. However, there are alternative mechanisms, besides having enormous specificity, for reducing the amount of protein required. For example, the whole genome is not accessible to the protein—due to the chromatin structure and other competing factors—reducing the amount of DNA that competes for binding to the protein. Another is that the specificity can be increased through cooperativity among factors. A single factor may only have a 10^6-fold increased affinity over nonspecific binding, but a complex of two or more proteins can have a much greater difference. Both of these mechanisms are used in organisms with large genomes, as is synthesis of many more copies of each TF than the number of regulatory sites.

Methods for Determining Specificity

Specificity is defined by the difference in binding affinity to different sequences. Thus, determining K_A to all possible binding sites would provide complete information about the specificity of a protein. Figure 5-5 parallels Figure 5-1 and contrasts the determination of specificity from that of affinity. Figure 5-1A illustrates a binding reaction between a protein and a single binding sequence, and Figure 5-1B shows how varying the concentration of protein allows determination of the affinity, defined as K_D, of the protein for that sequence. One could repeat that experiment for all

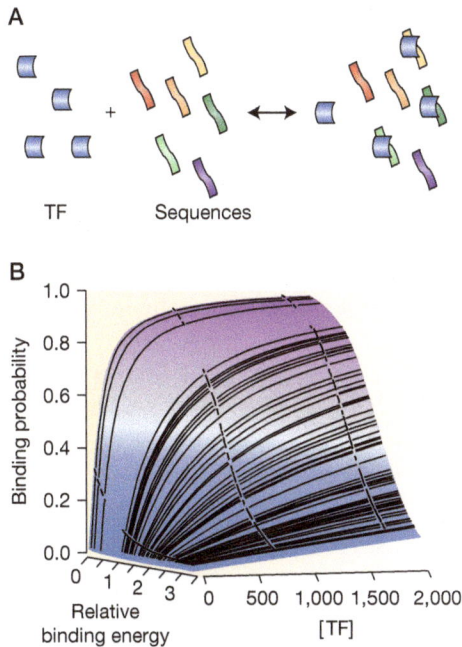

Figure 5-5. Specificity is the difference in binding affinities between different sequences. (A) DNA with different sequences (different colors) competes for binding to the same protein. (B) Plot of the binding probability (vertical axis) for all sequences (ordered by their energy on the left lower axis) for different concentrations of TF (lower right axis). For each individual sequence, the graph appears as in Figure 4-1B, but now, by comparing the binding curves of all sequences together, we see that TF has specificity because different sequences have different K_Ds. (Redrawn, with permission, from Stormo and Zhao 2010.)

possible sequences, but the number of possible binding sites of length L is 4^L—too many to be practically accomplished by one-at-a-time analysis for typical lengths of L (6–10 for monomeric binding proteins and longer for dimeric TFs). Figure 5-5A illustrates an experiment in which many different sequences compete to bind the same pool of protein. An assay that determines the fraction bound for each sequence in the mix, performed over varying protein concentrations, would generate a 3D (three-dimensional) curve, as in Figure 5-5B, allowing for the determination of K_D for each sequence simultaneously and providing a graphical representation of the specificity of the protein.

A variety of methods have been developed to determine specificities of TFs, and recent technological advances have allowed for HT assays that can determine specificity rapidly and accurately. Early methods relied on successive measurements of K_Ds of individual sequences. For example, Sarai and Takeda (1989) and Takeda et al. (1989) used a filter-binding assay (Fig. 5-2A) to measure the affinity of both Cro and λ repressor proteins to their shared 17-bp consensus sequence as well as to all 51 possible single-base variants to that sequence. If one assumes that the effects of multiple changes contribute independently (an additive energy model), that data would be sufficient to determine the affinity to any sequence above the nonspecific binding level. These investigators also measured the affinities to sites with multiple changes and showed that the additivity assumption is approximately, but not precisely, correct. However, even if additivity were to hold, each K_A measurement has some uncertainty, and predicting the K_As of multisite variants increases their uncertainty so that the accuracy of predicted K_As for many sequences will be low.

Because specificity depends on relative affinities of different sequences and not the absolute K_A of each sequence, one can both increase throughput and decrease uncertainty by measuring relative affinities of multiple sites simultaneously (Man and Stormo 2001). For example, Figure 5-6 shows binding to two different sequences, each labeled with a different fluorescent color, and running them on an EMSA gel (Fig. 5-6B). The first lane of the gel shows the total DNA in both the bound and unbound fractions, but lanes F1 and F2 are the same lane as viewed through different filters, so that each measures the amount of the two different DNAs independently. As shown in the figure equations, ratios of their K_As can be determined from the ratios of the fluorescent intensities in lanes F1 and F2. Concentration of protein is not important (and need not be varied) as long as it is not so low, or so high, that any of the four bands is not measurable. And although this figure illustrates the method for two DNA sequences, any number of sequences can be analyzed in a single lane of the gel as long as fluorescence signals can be well filtered so that independent intensities can be determined. Four-color assays, in which there is one color for each possible base at a position, can be used, allowing assay of all possible single-base variants for an L-long site to be determined in only L lanes of a gel. Multisite variants can also be tested to determine whether additivity holds, but it is still only practical to assay a small fraction of all potential binding sites for typical values of L.

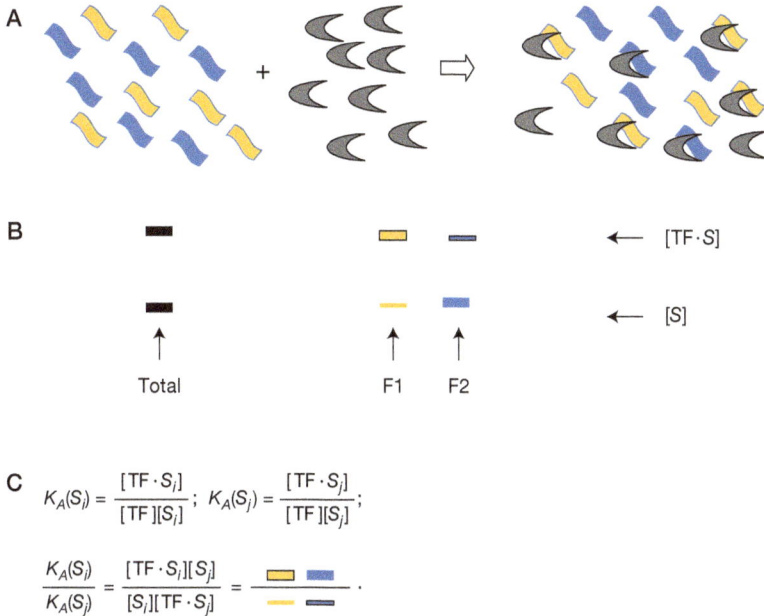

A

B

Total   F1   F2

$\leftarrow$ [TF·S]

$\leftarrow$ [S]

C

$$K_A(S_i) = \frac{[TF \cdot S_i]}{[TF][S_i]} \ ; \ K_A(S_j) = \frac{[TF \cdot S_j]}{[TF][S_j]} \ ;$$

$$\frac{K_A(S_i)}{K_A(S_j)} = \frac{[TF \cdot S_i][S_j]}{[S_i][TF \cdot S_j]} = \frac{\rule{1.2em}{0.4em} \ \rule{1.2em}{0.4em}}{\rule{1.2em}{0.4em} \ \rule{1.2em}{0.4em}} \ .$$

Figure 5-6. Determining relative affinity of two sequences in one EMSA lane. (A) Two different sequences are labeled with different fluorescent dyes and mixed with a protein to reach equilibrium. (B) Left is total DNA binding (both sequences combined) and center shows the same lane using two different filters that each allow only one labeled DNA to be observed. The upper band is the protein–DNA complex and the lower band is the free DNA (see Fig. 5-2B). (C) Equations for the association constant for each sequence are given (see Fig. 5-1B). The ratio of association constants does not depend on the free protein concentration and can be determined by the ratios of band intensities from the lanes F1 and F2.

SPR (Fig. 5-2C) can also be scaled up to assay several different sequences in parallel. By including on the surface an array of spots with different DNA sequences and scanning the laser across all of the spots, one can determine on and off rates for multiple different sequences for a given protein in parallel (Campbell and Kim 2007). So far, arrays have been used with up to a few hundred spots, but further increases seem possible. Other new technological advances make it practical to get much more data rapidly—data that can be used to estimate specificity of TF over all, or nearly all, possible binding sites. In the following paragraphs, several different methods are described.

Mitomi

Mechanically induced trapping of molecular interactions (MITOMI) uses microfluidic devices to perform hundreds to thousands of binding assays in parallel (Fig. 5-7A) (Maerkl and Quake 2007). TFs are synthesized within a chamber of the device using

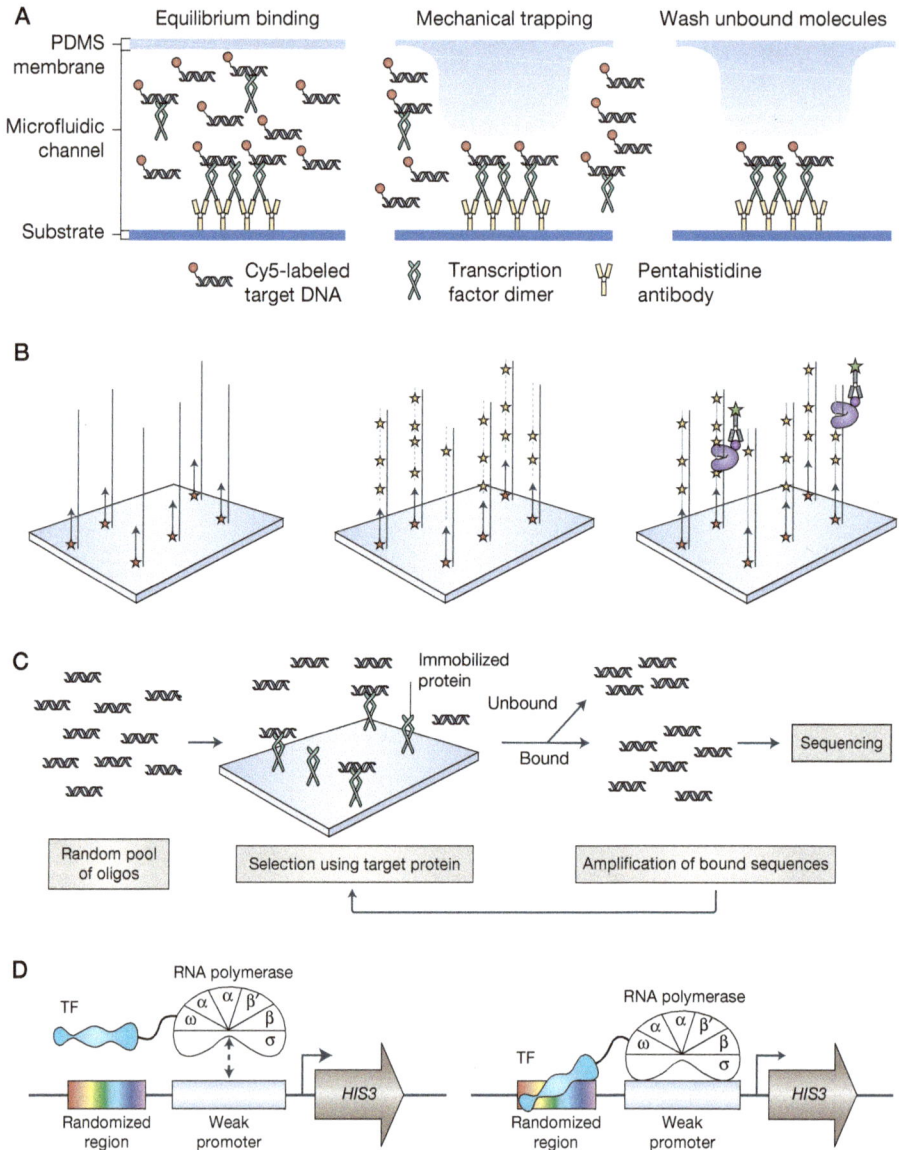

Figure 5-7. High-throughput methods for determining specificity of a DNA-binding protein. (A) Overview of the mechanically induced trapping of molecular interactions (MITOMI) device for binding affinity measurements. TF is bound to the surface by antibodies, and a fraction of the DNA binds to TF. Unbound DNA is expelled by washing so that the bound fraction can be measured. (PDMS) Polydimethylsiloxane. (Modified, with permission, from Maerkl and Quake 2007; © AAAS.) (B) Overview of universal protein-binding microarray (PBM) design and use. All possible 10-base-long sequences are included on the array (*left*). Primer-directed DNA synthesis creates dsDNA on the array (*middle*). (*Legend continues on following page.*)

in vitro transcription/translation mixtures. An antibody to the transcription attaches it to the bottom of the chamber. DNA can enter the chamber with the protein, and after equilibrium is reached, DNA that is not bound to protein is washed away. The DNA is labeled fluorescently, and the amount of DNA present before and after washing is measured to determine the fraction of DNA bound. The protein is also labeled with a different color so that comparisons among different chambers, which may have different amounts of protein, can be made. By performing these assays at several different DNA concentrations for each of many different DNA sequences, K_Ds can be estimated for each sequence. Current devices allow the determination of K_Ds for hundreds of DNA sequences rapidly and accurately, and by using longer DNA sequences that contain many potential binding sites, assays of all possible binding sites up to approximately length 8 are possible.

Protein-Binding Microarrays and Cognate Site Identifier Methods

Protein-binding microarrays (PBMs) (Berger et al. 2006) and cognate site identifier (CSI) methods (Puckett et al. 2007) are similar in that they both use dsDNA in a microarray format (Fig. 5-7B). PBM places single-strand DNA (ssDNA) on the array that is converted to dsDNA from a universal primer. CSI arrays ssDNA that can fold back on itself to produce dsDNA-binding sites. Current arrays allow for either method to contain all possible 10-long binding sites. TF is then added to the array, washed to remove nonspecific binding, and measured using a fluorescent antibody (or the protein itself can be fluorescently labeled). Comparing the fluorescence intensity at each spot on the array can lead to estimates of binding affinities to all possible binding sites up to approximately length 10.

SELEX

Systematic evolution of ligands by exponential enrichment (SELEX) (Tuerk and Gold 1990) uses a pool of random DNA (originally, RNA) sequences and a binding assay

---

Figure 5-7. (*Continued.*) Proteins (*right*, purple crescents) are bound, nonspecific binding is minimized by washing the array, and remaining protein is quantified with a fluorescent antibody. (Stars) Fluorescent groups. (Modified, with permission, from Berger et al. 2006 [© Macmillan Publishers Ltd.].) (C) Overview of HT-SELEX procedure. DNA molecules from the random library are exposed to TF; some sequences are bound, whereas others flow through. The bound fraction is sequenced to determine the probability of being bound. Rounds of amplification of bound sequences may be used (see main text). (Reprinted, with permission, from Stormo and Zhao 2010.) (D) Overview of bacterial one-hybrid (B1H) system. TF is fused to the ω subunit of RNA polymerase and a sequence from a randomized library is inserted upstream of the promoter of the *HIS3* gene (which encodes a component of the histidine biosynthesis pathway) (*left*). When TF binds to the randomized sequence, the *HIS3* promoter becomes more active (*right*) and this increases the growth rate of cells. (Modified, with permission, from Noyes et al. 2008 [© Oxford University Press].)

that selects binding sites based on their affinities (Fig. 5-7C). In the early days, when sequencing was limiting, multiple rounds of selection and amplification were performed and then a few of the remaining binding sites were sequenced. This generally identified the preferred binding sequence and gave information about the variability of binding preferences for different bases at different positions. New sequencing methods allow one to do a single round of selection, and by comparing the probabilities of all of the sequences in the initial library with the library after selection, to estimate the binding affinities to all possible sequences (Zhao et al. 2009; Zykovich et al. 2009). Further rounds of selection and sequencing can still be performed and provide additional information about relative affinities, especially for proteins with low specificity.

Bacterial One-Hybrid Methods

Bacterial one-hybrid (B1H) methods use a random library of binding sites driving the expression of a selectable gene in *E. coli* (Fig. 5-7D) (Meng et al. 2005). TF is expressed in cells under conditions in which cell growth depends on expression of the selectable protein (such as His3). Cells that make more of that protein will grow faster, up to some level at which is it no longer limiting. In the early days, a few colonies would be picked and sequenced and the preferred sequence and estimates of the contributions of different bases at different positions could be determined. Currently, one often sequences all of the cells on the plate (or even from growth in liquid media), obtains millions of reads for each selection, and from that estimates binding affinities for all possible binding sites.

Unified, Quantitative Definition of Specificity

Unlike affinity, there is no standard definition of specificity. In fact, two commonly used meanings exist. One, described in Chapter 4, is the ratio of the highest-affinity binding site to nonspecific binding. By this definition, a protein that has K_D of 10^{-9} M for its preferred binding site and a nonspecific binding affinity of 10^{-4} M has a specificity of 10^5. The other common meaning is related to how frequently binding sites are expected to occur by chance in a random sequence. This meaning is frequently applied to restriction enzymes, for instance, by referring to their binding-site lengths and degeneracies. For example, EcoRI (see Fig. 4-1) is referred to as a "six base cutter" (or simply "six cutter") because it cuts the site GAATTC. Similarly, AluI is called a "four cutter" because it cuts the site AGCT. Although from these examples specificity appears to refer to the length of the binding site, in reality it refers to the frequency of sites because the enzyme HinfI, which cuts the sites GANTC (where N can be any base), is also a four cutter; the frequency of its cut sites is the same as for AluI in that it will cut, on average, every 256 (4^4) bp. Similarly, HincII is a five cutter, even though its site contains six positions (GTYRAC) because Y ("T or C") and R ("A or G") each specify a "half base," meaning two of the four possible bases.

The frequency of its sites in a random sequence would be the same as that for an enzyme that recognized five specific bases, on average, every 1024 (4^5) bp.

Restriction enzymes (and some other proteins) are usually characterized as having "all-or-none" activity, in which sites are either cut or not (we ignore nicking activity here). This makes it convenient to define their specificity in bases, which combines the length of the restriction site and its degeneracy. TFs generally do not have such "all-or-none" activity, but rather the affinity for different sequences varies gradually between its preferred sequence and nonspecific binding, as illustrated in several of the examples in Chapter 4. There is a consistent approach to defining the specificity of a TF that takes into account its binding distribution across all possible sites, and that has, as its two extreme values, 0 for a nonspecific protein and L bases for a protein that binds to only one L-long sequence. We define a specific binding energy for each sequence and then compare the average specific binding energy to a nonspecific protein. If all different sequences, in equal concentration, are competing for binding to a single TF, the probability of it binding to sequence S_i with affinity $K_A(S_i)$ is

$$\Pr(S_i \text{ bound}) = K_A(S_i)/\Sigma_j K_A(S_j). \tag{5-4}$$

The sum in the denominator is over all possible sequences, which makes the denominator similar to the partition function Z described above, but it differs by not including the nonbound state or protein concentration. For a nonspecific binding protein that binds to a site of length L, $\Pr(S_i \text{ bound}) = 4^{-L}$ for every S_i; however, for a protein with specificity, the probability will vary depending on the sequence. A protein that only binds to one sequence has $\Pr(S_i \text{ bound}) = 1$ for that sequence and 0 for all other sequences. We define specific binding energy to sequence S_i as

$$E_{sp}(S_i) = -\log_4 [\Pr(S_i \text{ bound})] = -\log_4 K_A(S_i) + \log_4 \Sigma_j K_A(S_j). \tag{5-5}$$

For a nonspecific binding protein, $E_{sp}(S_i) = L$ for every sequence; thus, the average is also L. For a specific binding protein, the average specific binding energy is

$$\langle E_{sp} \rangle = \sum_i \Pr(S_i \text{ bound}) E_{sp}(S_i). \tag{5-6}$$

We can define the specificity of a protein as $L - \langle E_{sp} \rangle$. This will be 0 for a nonspecific binding protein and L for a protein that only binds to only one sequence of length L. All proteins that show differential affinity to different sequences will be somewhere in between, with the value corresponding to our intuition about specificity. For example, for HinfI and HincII enzymes described above, the value would be 4 and 5, even though the lengths of the binding sites are 5 and 6, respectively. But, in addition, for proteins that have a smooth distribution between the highest affinity and the lowest, as in Figure 5-5, this formula provides a consistent and useful definition of specificity. We use $\log_4$ so the results are in bases; if we used $\log_2$, the result would be in bits. We could also use standard energy units such as kT or kcal/mol but by describing a protein's specificity in bases, we can relate that to the amount needed for a regulatory system to function, as described earlier in this chapter.

Determining the specificity of a TF complements, and greatly extends, the information available from the structural studies described in Chapter 4. Although it would not be feasible to determine structures of complexes for all possible binding sites, determining their affinities is made possible by new technologies. Combining principles observed in structural studies with large data sets available from binding assays is valuable for the bioinformatics analyses described in Chapter 7 and then for use in systems biology and synthetic biology, as described in Chapter 9. In Chapter 6, we delve more deeply into thermodynamics and kinetics of binding and the ways in which different types of interactions contribute to overall binding free energy.

REFERENCES

Berger MF, Philippakis AA, Qureshi AM, He FS, Estep PW, 3rd, Bulyk ML. 2006. Compact, universal DNA microarrays to comprehensively determine transcription-factor binding site specificities. *Nat Biotechnol* **24**: 1429–1435.

Campbell CT, Kim G. 2007. SPR microscopy and its applications to high-throughput analyses of biomolecular binding events and their kinetics. *Biomaterials* **28**: 2380–2392.

Maerkl SJ, Quake SR. 2007. A systems approach to measuring the binding energy landscapes of transcription factors. *Science* **315**: 233–237.

Man T-K, Stormo GD. 2010. Non-independence of Mnt repressor-operator interaction determined by a new quantitative multiple fluorescence relative affinity (QuMFRA) assay. *Nucl Acids Res* **29**: 2471–2478.

Meng X, Brodsky MH, Wolfe SA. 2005. A bacterial one-hybrid system for determining the DNA-binding specificity of transcription factors. *Nat Biotechnol* **23**: 988–994.

Noyes MB, Meng X, Wakabayashi A, Sinha S, Brodsky MH, Wolfe SA. 2008. A systematic characterization of factors that regulate *Drosophila* segmentation via a bacterial one-hybrid system. *Nucleic Acids Res* **36**: 2547–2560.

Puckett JW, Muzikar KA, Tietjen J, Warren CL, Ansari AZ, Dervan PB. 2007. Quantitative microarray profiling of DNA-binding molecules. *J Am Chem Soc* **129**: 12310–12319.

Sarai A, Takeda Y. 1989. λ Repressor recognizes the approximately 2-fold symmetric half-operator sequences asymmetrically. *Proc Natl Acad Sci* **86**: 6513–6517.

Stormo GD, Zhao Y. 2010. Determining the specificity of protein-DNA interactions. *Nat Rev Genet* **11**: 751–760.

Takeda Y, Sarai A, Rivera VM. 1989. Analysis of the sequence-specific interactions between Cro repressor and operator DNA by systematic base substitution experiments. *Proc Natl Acad Sci* **86**: 439–443.

Tuerk C, Gold L. 1990. Systematic evolution of ligands by exponential enrichment: RNA ligands to bacteriophage T4 DNA polymerase. *Science* **249**: 505–510.

Zhao Y, Granas D, Stormo GD. 2009. Inferring binding energies from selected binding sites. *PLoS Comput Biol* **5**: e1000590.

Zykovich A, Korf I, Segal DJ. 2009. Bind-n-Seq: High-throughput analysis of in vitro protein-DNA interactions using massively parallel sequencing. *Nucleic Acids Res* **37:** e151.

FURTHER READING

Geertz M, Maerkl SJ. 2010. Experimental strategies for studying transcription factor-DNA binding specificities. *Brief Funct Genomics* **9:** 362–373.

Helwa R, Hoheisel JD. 2010. Analysis of DNA-protein interactions: From nitrocellulose filter binding assays to microarray studies. *Anal Bioanal Chem* **398:** 2551–2561.

Oda M, Nakamura H. 2000. Thermodynamic and kinetic analyses for understanding sequence-specific DNA recognition. *Genes Cells* **5:** 319–326.

CHAPTER 6

Energetics and Kinetics of Binding

T HE FORCES THAT DRIVE THE FORMATION OF THE PROTEIN–DNA complex can be described in several different and complementary ways. At the level of fundamental physics, the free energy of binding is a combination of enthalpy, the internal energy of the bonds formed in the complex, and entropy, related to the number of microstates available to the system. One can also consider the contributions of individual atomic interactions between the DNA and protein, based on structural information or detailed experimental characterization, and categorize them into different classes. For example, it is relatively straightforward to separate electrostatic from nonelectrostatic contributions. It is also useful to separate interactions into those that contribute to the sequence-specific or nonspecific binding affinity. The sequence-specific contributions can be further divided into direct readout of the base sequence and readout of the sequence-dependent shape of the DNA. Computational analyses based on the structures of complexes, as described in Chapters 3 and 4, and biophysical forces, described in this chapter, can provide further insight into the thermodynamics of the interaction and aid in the design of proteins with novel specificities. Finally, the kinetics of the binding reaction can be analyzed. There is an inherent conflict between rapid location of the regulatory sites and the formation of highly stable complexes. Proteins have evolved a balance between those opposing traits that allows for regulatory systems to function efficiently and be tunable to attain specifically desirable characteristics for individual systems.

THERMODYNAMICS OF TRANSCRIPTION FACTOR BINDING

Free Energy of Binding

In biological systems, which are generally at constant pressure, the energetics of protein–DNA binding is best described by the Gibbs free energy:

$$G = H - TS, \tag{1}$$

where H is the enthalpy, S the entropy, and T the temperature (in degrees Kelvin). At equilibrium, the free energy of the system is at a minimum and at constant temperature,

$$\Delta G = \Delta H - T\Delta S \tag{2}$$

describes how far the system is from equilibrium and in which direction it will move spontaneously, either to form more or less of the complex. When $\Delta G < 0$, the system will spontaneously form more of the complex, and when $\Delta G > 0$, the existing concentration of the complex will decrease to attain equilibrium (see Box 6-1). Figure 5-1 defined the ratio of concentrations of the reactants (the unbound protein and DNA independently) to the concentration of the complex as the dissociation constant

$$K_D = \frac{[TF]_{eq}[S]_{eq}}{[TF \cdot S]_{eq}}, \tag{3}$$

where eq refers to their concentrations at equilibrium. The free energy of the system at any nonequilibrium state of the system, x, can be calculated from the concentrations of reactants and products in that state, from

$$\Delta G_x = RT\left[\ln K_D - \ln \frac{[TF]_x[S]_x}{[TF \cdot S]_x}\right], \tag{4}$$

Box 6-1. Calculation of ΔG in Different Conditions, Including $\Delta G°$ for the Standard Condition of 1 M in Reactants and Products

---

In this example, TF (blue crescents, Box Fig. 6-1) has a K_D of 0.0357 (1/28) for binding to the DNA sequence S (gold ribbons). On the left side, the top panel has TF and S in separate compartments, each at a concentration of 4 M. Each separate compartment is at equilibrium, so that $\Delta G = 0$. When the barrier between the compartments is removed (right side), TF and S concentrations drop to 2 M and there is initially no complex, so $\Delta G = -\infty$. In the third panel (middle, right), half of the TF and S have formed complexes, so that everything is at 1 M, the standard condition at which $\Delta G° = -RT \ln K_D = -3.33RT$. R (the "gas constant") = 1.985 cal/K°mol (degrees Kelvin times moles). At "body temperature" 310°K (37°C), $RT \approx 0.62$ kcal/mol, which is a commonly used unit of free energy in biology. Thus, for this TF and S, $\Delta G° = -2.07$ kcal/mol. In the fourth panel (middle, left), the system has reached equilibrium with [TF] = [S] = 0.25 M and [TF·S] = 1.75 M and $\Delta G = 0$. In the fifth panel (lower left), the system has been diluted by increasing the volume by a factor of 4.66 so that the concentrations of TF and S are instantly at 0.054 M and the complex is at 0.375 M. This ratio is higher than equilibrium, so $\Delta G = 0.96$ kcal/mol. In the last panel, the concentrations of TF and S have doubled to 0.108, the concentration of the complex has decreased to 0.321 (6/7 of its previous value), and the system has again reached equilibrium with $\Delta G = 0$. Box Figure 6-2 graphs ΔG for the different panels.

Box 6-1. (Continued.)

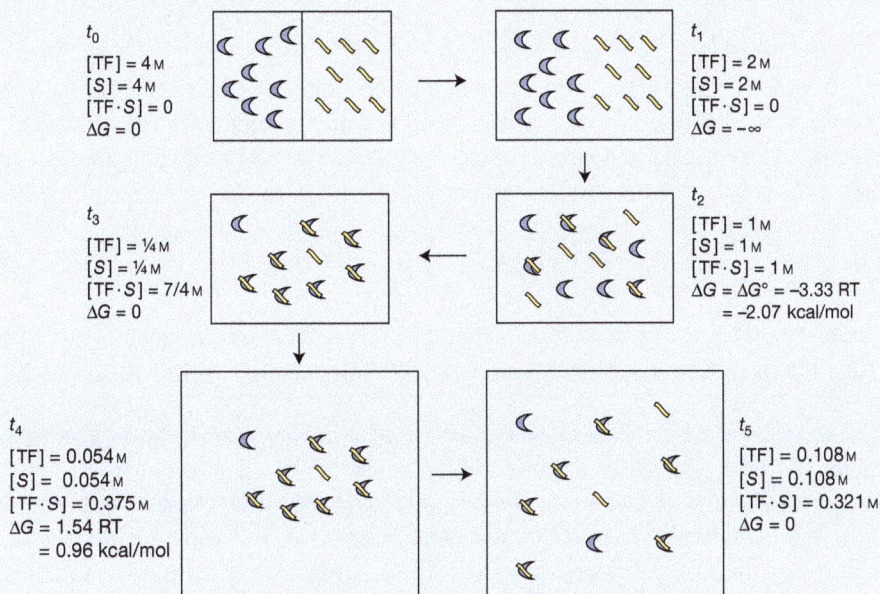

t_0
[TF] = 4 M
[S] = 4 M
[TF·S] = 0
$\Delta G = 0$

t_1
[TF] = 2 M
[S] = 2 M
[TF·S] = 0
$\Delta G = -\infty$

t_3
[TF] = ¼ M
[S] = ¼ M
[TF·S] = 7/4 M
$\Delta G = 0$

t_2
[TF] = 1 M
[S] = 1 M
[TF·S] = 1 M
$\Delta G = \Delta G° = -3.33$ RT
       = −2.07 kcal/mol

t_4
[TF] = 0.054 M
[S] = 0.054 M
[TF·S] = 0.375 M
$\Delta G = 1.54$ RT
       = 0.96 kcal/mol

t_5
[TF] = 0.108 M
[S] = 0.108 M
[TF·S] = 0.321 M
$\Delta G = 0$

Box Figure 6-1. Calculation of ΔG under different conditions.

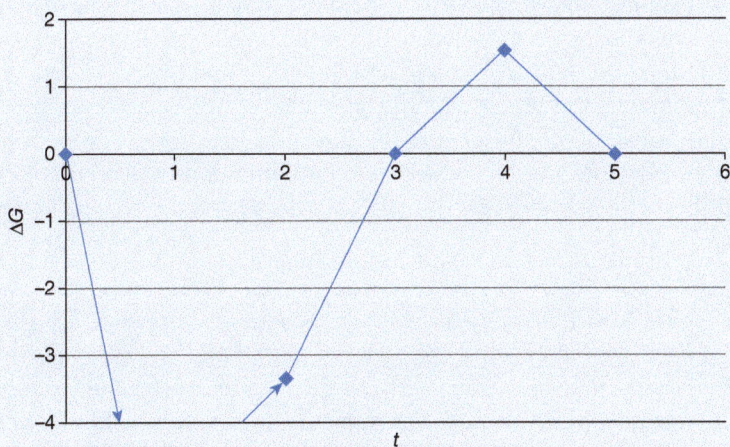

Box Figure 6-2. Graph of ΔG values for the panels in Box Figure 6-1.

where R is the "gas constant" (see Box 6-1) and T is again the temperature. The "standard state" is defined as 1 M in the reactants and products, for which the free energy is

$$\Delta G^{\circ} = RT \ln K_D = -RT \ln K_A. \tag{5}$$

The standard free energy for a particular protein–DNA complex is useful to know because of its relationship to the dissociation (and association) constant, and the free energy of any particular state can be determined from the concentrations of reactants and products in that state and the standard free energy by

$$\Delta G_x = \Delta G^{\circ} - RT \ln \frac{[TF]_x[S]_x}{[TF \cdot S]_x}. \tag{6}$$

The sign of ΔG_x indicates the direction that the reactors go, to make more or less of the complex. Box 6-1 provides a simple example of the free energy in different states of a system.

The enthalpy of an interaction can be measured using isothermal titration calorimetry (ITC), in which the amount of heat added or removed from a reaction to maintain constant temperature is determined (Fig. 6-1). With precision instruments, this method is capable of measuring very small changes in heat released, or adsorbed, by the formation of the bonds created in the protein–DNA complex. One might believe that binding would always release heat, with the complex in a lower enthalpy state than the protein and DNA independently, but that is not always true. For example, the formation of hydrogen bonds in the complex is at the expense of hydrogen bonds formed separately by the protein and DNA with water in the surrounding solvent. Similarly, electrostatic interactions between the negatively charged phosphates of the DNA backbone and charged amino acids on the protein generally replace similar interactions with the ions in the solvent. The net change in the internal energy of all of the bonds formed between the protein and the DNA, and the bonds lost among the solvent, the protein, and the DNA independently may be small or even positive. If positive, the formation of the complex is driven by an increase in entropy. Given the ΔG of the reaction, which can be determined from the equilibrium binding constant, and the ΔH determined by ITC, one can determine the change in entropy by

$$T\Delta S = \Delta H - \Delta G. \tag{7}$$

How is formation of a complex driven by increasing entropy? At first glance, one expects the entropy of the complex to be lower than that of the independent protein and DNA reactants. In the complex, both components are constrained to move about in solution together, which reduces the number of microstates allowed when compared to the number of states when they move independently. But missing from that consideration is the release of water and ions from each of the reactants. Both protein and DNA are generally associated with ions that neutralize their charges (not all proteins have a net charge, but most transcription factors have some charged

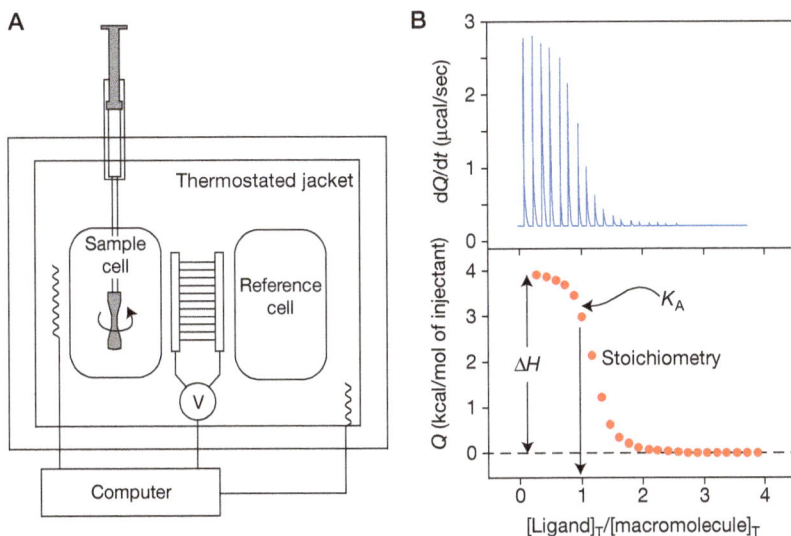

Figure 6-1. (A) Basic schematic illustration of the ITC instrument, showing the two cells (sample and reference) surrounded by the thermostated jacket, the injection syringe that also works as a stirring device, and the computer-controlled thermostatic and feedback systems (using Peltier and resistor devices as sensor and actuator subsystems). (B) Example of a typical ITC experiment. (*Top panel*) The sequence of peaks, each one corresponding to each injection of the solution in the syringe. The monitored signal is the additional thermal power needed to be supplied or removed at any time to maintain a constant temperature in the sample cell and as close as possible to the reference cell temperature. This example corresponds to endothermic binding. (*Bottom panel*) The integrated heat plot. The areas under each peak, calculated and normalized per mole of ligand injected in each injection, are plotted against the molar ratio (quotient of the total concentrations of ligand and macromolecule in the sample cell). From this plot, and applying the appropriate model, the thermodynamic parameters of the binding can be obtained: binding affinity, binding enthalpy, and stoichiometry. (Reprinted, with permission, from Velázquez-Campoy et al. 2004 [©2004 John Wiley & Sons, Inc.].)

residues on their surface). In addition, water molecules readily form hydrogen bonds with polar atoms of DNA and protein and also tend to be somewhat ordered around nonpolar atoms. Those associated water and ion molecules are in a semistructured state, in which they are constrained in their orientation about the protein and DNA. Of course, they are not held in rigid complexes and can be readily exchanged with the bulk solvent, but their movement and orientation are partially constrained. When the complex forms, many of those associated water molecules and ions are released into bulk solvent where they are much freer in their movements, thereby increasing the total entropy of the system. For some protein–DNA complexes, this is the primary driving force for complex formation, overcoming a positive change in enthalpy; however, individual proteins fall on a broad spectrum. Although $\Delta G°$ values for transcription factors are always negative, and generally in the range of -8 to -14

kcal/mol for sequence-specific binding (K_D values of 10^{-6} to 10^{-11} M), the contributions of ΔS and ΔH can be quite variable. Table 6-1 lists measured values of $\Delta G°$ and its separation into ΔH and $T\Delta S$ for two restriction enzymes and several transcription factors from multiple different classes. The specific values depend significantly on the DNA site bound by each protein as well as the conditions, especially salt concentration (see below) and temperature; thus, experiments performed under different conditions often report different values. Those reported in Table 6-1 are for the preferred binding site at reasonably high salt (usually 100 mM NaCl or KCl or more) and at ~20°C (or the reported temperature nearest that value). The table shows that although these DNA-binding proteins have a fairly small range of $\Delta G°$ values, they can vary enormously in enthalpy and entropy contributions. For many proteins, both are favorable, showing a negative ΔH and a positive $T\Delta S$. In other cases, one of the values is near 0 and the free energy is determined by the other value. For example, both the Trp repressor and c-Myb have small values of $T\Delta S$, whereas for Antp and Sox-5, it is the ΔH value that is small (Sox-5 even has small positive ΔH under somewhat different conditions [Privalov et al. 1999]). And, in a few cases, one of the values is highly unfavorable so that the other value must be highly favorable to attain the $\Delta G°$ needed. For example, both cAMP receptor protein (CRP) and Sry have large unfavorable ΔH values that are compensated for with an even larger favorable $T\Delta S$. In both cases, the large positive ΔH values are due to a large bend in the DNA, which is

Table 6-1. Standard free energies, enthalpies, and entropies for several protein–DNA interactions

Transcription factor	Class	$\Delta G°$	$\Delta H°$	$T\Delta S°$	Reference
EcoRI	Restriction enzyme	−14.6	−9.7	4.9	Jen-Jacobson et al. 2000
BamHI	Restriction enzyme	−12.7	−16.6	−4.0	Jen-Jacobson et al. 2000
CRP	HTH	−9.2	20.1	29.4	Shi et al. 1999
λ Repressor	HTH	−11.9	−28.4	−16.5	Merabet and Ackers 1995
Trp repressor	HTH	−12.1	−12.8	−0.7	Ladbury et al. 1994
C-Myb	HTH	−12.1	−12.5	−0.4	Oda et al. 1998
Mbp1	Winged HTH	−10.6	−9.6	1.0	Deleeuw et al. 2008
Gcn4	bZIP	−11.8	−32.6	−20.9	Dragan et al. 2004a
Jun-Fox	bZIP	−9.1	−35.1	−25.9	Seldeen et al. 2009
Mash-1	bHLH	−10.1	−24.2	−14.2	Kunne et al. 1998
Antp	Homeodomain	−12.9	−2.9	10.0	Dragan et al. 2006
Engrailed	Homeodomain	−10.8	−4.5	6.2	Dragan et al. 2006
YY1	Zinc finger	−8.6	−7.6	1.0	Houbaviy and Burley 2001
TFIIIA	Zinc finger	−9.6	−4.1	5.5	Liggins and Privalov 2000
MetJ	B-ribbon	−7.9	−4.1	3.9	Hyre and Spicer 1995
Sry	HMG	−14.8	8.4	23.2	Dragan et al. 2004b
Sox-5	HMG	−11.2	−3.0	8.2	Privalov et al. 1999
TBP from Archaea	TATA-binding protein	−7.2	26.3	33.5	Bergqvist et al. 2001

In each case, the DNA is the preferred binding site for the protein. All values are in kcal/mol, with temperature usually at or near 20°C. CRP, cAMP receptor protein; HTH, helix turn helix; bZIP, basic-region leucine zipper; bHLH, basic-region helix loop helix; HMG, high mobility group; TBP, TATA-binding protein.

energetically costly, after binding. The TATA-binding protein (TBP), here from Archae-bacteria, is an extreme example of a very large induced bend on binding to the minor groove (see Fig. 3-7A). In other cases, $T\Delta S$ is very unfavorable (negative) and is compensated for by a large negative ΔH. Basic-region leucine zipper (bZIP) and basic-region helix-loop-helix (bHLH) proteins fall into that class, and in each case, this is due to that fact that the DNA-binding domain of the protein is largely unstructured before binding, which requires a large decrease in entropy from simultaneous folding and binding. In addition, one can generalize that protein binding to the minor groove often has $\Delta H > 0$ and is driven by large increases in entropy due to the release of highly ordered water lining the minor groove, especially for AT-rich sequences. Binding in the major groove generally has $\Delta H < 0$, and entropy is smaller than for the minor groove (it is even negative in some cases) (Privalov et al. 2007).

Heat Capacity Changes

The heat capacity change after formation of the complex is also informative as to the nature of the interaction. Heat capacity, at constant pressure, is

$$C_P = \frac{dH}{dT} \tag{8}$$

or the change in enthalpy divided by the change in temperature. Heat capacity, the amount of heat required to raise the temperature by $1\,^{\circ}\text{C}$ (or K), measures how much energy can be taken up by the molecules in solution without contributing to their kinetic energy, which is measured by the temperature. This includes intramolecular vibrational and rotational energies as well as intermolecular interactions between the protein or DNA and the solvent molecules organized around them. Both hydrophilic and hydrophobic regions of proteins tend to be surrounded by semiorganized water, which will adsorb heat in vibrational modes, and DNA also organizes adjacent water. The change in heat capacity ΔC_P after formation of the complex is

$$\Delta C_P = \frac{d\Delta H}{dT} \tag{9}$$

and can provide insights into the molecular processes that occur. For example, when a protein unfolds into a random structure, the heat capacity generally increases in proportion to the surface area that becomes exposed to the solvent. Heat capacity after unfolding can also increase because of additional vibrational and rotational modes in the less compact, unfolded form, but often it is the increased surface area, and its hydration by the solvent, that is the primary source of the increased heat capacity. The dehydration of hydrophobic surfaces when proteins fold into their native state is called the "hydrophobic effect" and is usually a significant contribution to the free energy of the folded protein.

For the study of protein–DNA interactions, the change in heat capacity of the complex compared to the independent protein and DNA can provide insights about the binding mechanism. Nonspecific binding, which is primarily to the DNA backbone, typically has a small ΔC_P. This is consistent with the model of nonspecific binding having a small ΔH, being driven by an increase of entropy through the release of ions interacting with the DNA and protein, and involving only small changes in accessible surface area. Specific binding to the major groove generally has a large negative ΔC_P due to the decrease in solvent-accessible surface in both the protein and DNA. The negative ΔC_P can be very large if the protein is unfolded in the unbound state and becomes folded after binding, such as occurs for the bZIP protein Gcn4 (see Fig. 3-5A) (Weiss et al. 1990). Proteins binding specifically in the minor groove usually have smaller values of ΔC_P than do major groove interactions, and they can be either positive or negative depending on details of the interaction. If the interaction is largely electrostatic, ΔC_P is likely to be positive. For interactions with the minor groove faces of the base pairs, especially for AT-rich sequences that have well-ordered water molecules lining the minor groove, ΔC_P is generally negative (Privalov et al. 2007).

Determining ΔG, ΔH, ΔS, and ΔC_P for a protein–DNA interaction can provide valuable information about the molecular mechanisms that drive complex formation. Different combinations of enthalpy and entropy are used by different TFs, but the range of free energy is relatively small. Changes in heat capacity can help to further elucidate the mechanism of binding. Figure 6-2 diagrams many of the possible interactions between protein and DNA and their contributions to the free energy and heat capacity.

Molecular Contributions to Complex Formation

Most of the types of bonds that contribute to the enthalpy of the interaction are shown in the figures in Chapter 4. Interactions that govern specificity are largely the hydrogen bonds between polar amino acids and base pairs, either in the major or minor grooves of the double helix, and nonelectrostatic and non-hydrogen-bond interactions of van der Waals' contacts (weak interactions among atoms in close proximity). The interactions that govern nonspecific interactions, contributing significant free energy to the overall affinity but not distinguishing among different sequences, include direct electrostatic interactions between the negatively charged phosphate backbone and positively charged amino acids, especially Arg and Lys, and also include hydrogen bonds to the sugar–phosphate backbone by polar amino acids. Interactions to the DNA backbone can also contribute to sequence specificity, especially in cases in which the DNA structure in the protein–DNA complex differs from the standard B-form helix because the energy of deformation and the propensity for different shapes is sequence dependent. Finally, there are indirect contacts between amino acids and the DNA mediated by an intervening water molecule, several examples

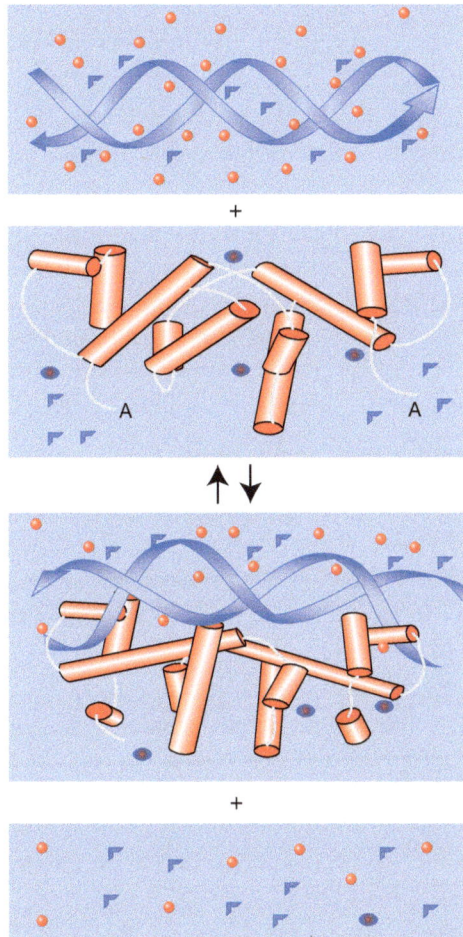

Figure 6-2. (*Top panel*) DNA has closely associated ions (red balls) and some highly ordered water molecules (blue arrowheads). (*Second panel*) Protein also has some ions (red/blue balls) and some highly ordered water molecules. The cylinders correspond to α-helical regions of the proteins. The segment marked "A" is unstructured when the protein is free in solution and has ordered water around it. (*Third panel*) In the protein–DNA complex, the DNA structure is slightly altered from the solution structure and the unordered segment of the protein marked "A" above is now an α helix. Some ions and ordered water molecules remain with the complex, but many of them are released into the bulk solvent (*bottom panel*). The contributions to binding free energy include the release of the ions and ordered water molecules, both of which increase the entropy of the system, whereas the new α helix will decrease the entropy somewhat. Distortion of the DNA will increase the enthalpy, but the bonds between the protein and DNA may compensate so that the overall enthalpy is lower. The total free energy change is the sum of all the entropy and enthalpy contributions. The change in heat capacity is probably negative because of the reduced solvent-accessible surface area at the protein–DNA interface and because of the new α-helical segment. (Modified, with permission, from Jayaram and Jain 2004 [©2004 *Annual Reviews*].)

of which are diagrammed in Chapter 4. Together, all of the contacts between protein and DNA contribute to the enthalpy of the complex for both specific and nonspecific binding energies (Fig. 6-2).

Electrostatic and Nonelectrostatic Contributions

The net release of ions after binding can be determined by measuring the salt dependence of the binding affinity. In Chapter 4, the association constant is defined only in terms of the protein, DNA, and complex concentrations because we implicitly assume a standard condition, typically one that is approximately physiological. Instead of simply using the notation K_A, we could use a more detailed notation $K_A(TF,S_i)_x$ to indicate that we are referring to the association constant of the TF to a particular sequence S_i under the specific conditions x, which includes the temperature, pH, all of the ion concentrations, and any other components of the reaction that may influence the measured value of K_A. In fact, the association constant and binding free energy depend very much on the conditions; in particular, the ionic concentrations. We can rewrite the reaction with the specific inclusion of ions that are tightly associated with the DNA and that are released on binding as

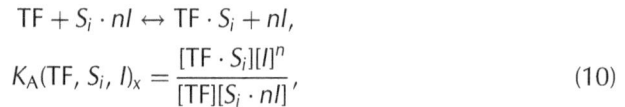

$$TF + S_i \cdot nl \leftrightarrow TF \cdot S_i + nl,$$

$$K_A(TF, S_i, l)_x = \frac{[TF \cdot S_i][l]^n}{[TF][S_i \cdot nl]}, \tag{10}$$

where l represents ions complexed with the DNA that are released after binding and now x refers to all of the other components of the reaction except for TF, S_i, and l that are explicitly included as variables. This is still a simplification because ions can also be associated with the protein, but generally their contributions are smaller than the release of ions from the polyanionic DNA. And there may be different types of ions that would need to be considered independently in a detailed analysis. In vitro binding studies have control over the ions and their concentrations, and varying them can determine the electrostatic contribution to binding. In this simplified analysis,

$$K_A(TF, S_i, l)_x = K_A(TF, S_i)_x[l]^n, \tag{11}$$

where the conditions x are the same in both cases, but on the left, the ion concentration is included explicitly as one of the variables (Eq. 10), and on the right, we are using the definition of K_A from Chapter 5. Because the number of ions released is the exponent of the ion concentration, the association constant can be enormously sensitive to that concentration (depending on the value of n). This means that we can determine the number of ions released after binding by measuring $K_A(TF, S_i)$ under different ion concentrations:

$$\frac{d \ln K_A(TF, S_i)_x}{d \ln [l]} = -n, \tag{12}$$

where the slope of the log-log plot of K_A versus $[I]$ is the value of n. The valency of the ion matters. For example, if there are n phosphates whose interacting ions are released after binding, that would require n Na^+ ions but only $n/2$ Mg^{++} ions because of their double charge (each Mg^{++} can neutralize two phosphates). In vivo, it is generally Mg^{++} that is bound to the DNA, but in vitro binding assays often use Na^+ because the greater sensitivity (n vs. $n/2$) makes it easier to measure accurately. ITC can also be performed under different salt conditions to determine how much the ion release contributes to the enthalpy and entropy of the interaction. In general, electrostatic interactions contribute primarily to increases in entropy because of the large number of more unconstrained ions in the bound state. Different proteins have different numbers of ions released and in some cases the number varies between the nonspecific and specific binding modes of the protein. For example, the Lac repressor releases ~11 ions when bound nonspecifically but only about six when bound specifically to the operator DNA due to a different orientation of the protein when it interacts with the base pairs directly (Chapter 4) (Kalodimos et al. 2004).

Specific and Nonspecific Contributions

Although the affinity is much lower for nonspecific binding—typically, the K_{AS} are reduced by 10^4–10^7-fold—they are often high enough to be measured by standard methods, such as those described in Chapter 5 and by ITC. As described above, the electrostatic contribution to binding is primarily entropic and nonspecific. The enthalpic contributions are primarily to specific binding and most often involve direct contact with base pairs in the major groove. The base pairs within a binding site that contribute to specific binding can best be determined by assaying the affinity to sequence variations. If all base pairs at a position in the binding site have the same affinity, that position does not contribute to the specificity, although it may make important contributions to affinity (for example, if the contact is to the DNA backbone). One can also determine the contributions of individual components of specific base pairs to the binding affinity by using synthetic DNA with modified bases. For example, U differs from T only in that it is missing the methyl group in the major groove (see Fig. 2-1), so DNA that is synthesized with U instead of T at particular positions can be used to determine the contribution of those methyl groups to binding. Other modified bases, such as purine analogs, can also be substituted into binding sites to probe the contributions of specific components of individual bases. One can also identify particular phosphates that contribute to binding by chemical modifications. For example, ethylnitrosourea will ethylate phosphates and reduce the binding affinity for those that are contacted directly by the protein. Using light treatment of DNA with ethylnitrosourea, such that individual binding sites have at most one modified phosphate, and then separating bound from unbound DNA, the phosphates that are important for binding will be underethylated compared to those that do not interact with the protein. The ethylated positions can be identified because of their increased sensitivity to cleavage by NaOH.

Direct and Indirect Readout

Direct and indirect readout both refer to contributions to sequence-specific binding. In direct readout, contributions come directly from interactions with base pairs, such as through hydrogen bonds and van der Waals interactions, which depend on the sequence. Many examples of those are shown in Chapter 4. Indirect readout refers to interactions that depend on the sequence indirectly through its effects on the shape of the DNA. These can be determined by comparing mutational studies with structural information, in which particular bases are identified that contribute to the specificity, but there are no direct contacts apparent from the structure. Indirect contributions can occur because some sequences are more likely to be intrinsically bent, and if that contour better matches the surface of the protein, such as is the case with CRP (Fig. 1-1), the interaction can occur with less strain on either the protein or DNA. It can also occur because the widths of the major and minor grooves vary, depending on the sequence, and that can influence binding affinity. For example, AT-rich sequences tend to have narrower minor grooves than standard B-form DNA, and that leads to a higher density of negative charge, which interacts favorably with Arg residues. The definitions of direct and indirect binding are mutually exclusive, so an individual base pair can be one or the other. But there are many cases in which base pairs influence the structure of the DNA to make it more favorable for binding but also interact directly with the protein. Therefore, it has been suggested that a better classification is between base readout for direct contacts and shape readout for sequence-specific contributions to binding that depends on the structure of the DNA. Individual bases may contribute in both ways to such a classification of sequence-specific contributions to the interaction (Rohs et al. 2010).

Computational Modeling

Protein structure analysis and modeling represents one of the earliest applications of computers to biological problems. An active area of research has been to develop computational methods to accurately predict protein structures from their sequences, especially after the demonstration that the information for proper protein folding was contained within the protein sequence itself; many purified proteins could be denatured and then renatured to full activity in vitro. Achieving that goal has been very challenging, but methods have been improving due to a combination of better algorithms and increased computer speed, which allows for a more thorough exploration of the possible conformations. A more tractable problem is to start with a known structure for a protein and study its dynamics computationally or study its interactions with other molecules whose structure is also known—either other proteins or small molecules such as substrates, ligands, drugs, etc. The study of protein–DNA interactions fits in this category and has been an active area of research in recent years.

There are two primary approaches used to analyze the dynamics and energetics of structures computationally. Molecular dynamics (MD) simulations start with a structure and calculate the force on each atom based on a collection of force-field parameters, such as electrostatic attractions and repulsions and van der Waals interactions. At each time step, the atoms move in accordance with those forces, the forces are recalculated based on the new positions, and the atoms move again. Tracking the movements for enough steps to simulate nanoseconds to microseconds (which may take hours to days of real computer time, depending on the complexity of the molecules and the details of the algorithm) provides a view of the dynamics of the molecules, how their shapes change through time, and what fraction of the time is spent in different conformations.

Monte Carlo (MC) methods, also referred to as Markov chain Monte Carlo (MCMC), use a different approach but are also based on the same or similar force-field parameters. Again, starting with a structure, its internal energy is calculated, and then some random small displacement of the molecule is made and the energy recalculated. If the energy decreases, the move is accepted; if the energy increases, the move is accepted with a probability equal to $e^{-\Delta G/RT}$ and otherwise rejected (back to the previous conformation). This means that the molecule will tend to move toward lower-energy conformations, but it will also make "uphill" moves with a probability related to how much the energy increases. If the simulation is run long enough, it will provide an overview of the energy landscape, determining the free energies of the various conformations.

For both MD and MC, a number of considerations affect their accuracies as well as the computational resources required. Force fields contain parameters for electrostatic and van der Waals interactions among atoms as well as parameters for covalent bond distances, angles, and torsions. More complex parameters, for instance, accounting for induced polarization of atoms, can be added at increased computational expense. AMBER and CHARMM are two commonly used force fields that have been in use for many years but are also regularly updated, and several other force-field parameter sets are also commonly used (Ponder and Case 2003). In addition, the treatment of the solvent is a critical aspect of each approach. Explicit solvent methods include individual water molecules and ions in the calculation, and their moves are included in the dynamics, which significantly increases the time required for the simulations. Because of that time cost, most methods use an implicit solvent approach that replaces explicit solvent molecules with a dielectric continuum, either fixed or variable. Although much simpler, and therefore faster, the implicit solvent approach can account for many of the properties of interactions between proteins and nucleic acids and the solvent (Feig and Brooks 2004).

Computational modeling is valuable both for explaining and understanding the energetics of protein–DNA interactions and also for predicting and designing novel interactions. Some contributions to the binding energy are not easily distinguished by experimental methods but can be studied using computational simulations. For example, a protein–DNA complex may show bending of the DNA compared to the standard B-form structure. One wants to know whether that bending is intrinsic to the particular DNA sequence or whether it is induced in the DNA to better accommodate interactions with the protein, and, if it is induced, what the energetic cost is. Although it is possible to determine the structure of the DNA in the absence of protein to know if it is inherently bent, that has only been performed for a few different DNA sequences and still would not reveal the energetic difference between different conformations. Computational approaches can be used to estimate those differences for any DNA sequence and, in fact, obtain energetic parameters for all of the possible deformations of di-, tri-, and tetranucleotide sequences that probably constitute all of the essential parameters for sequence-dependent conformations (Fujii et al. 2007) (see Chapter 2). Such studies have revealed that for some DNA sites that are in non-B-form conformations in the complex, such as being bent or with narrowed minor grooves, that is the minimum energy conformation, whereas for others the bound conformation is induced by the protein interaction, and the energetic cost of that can be estimated (Rohs et al. 2009).

To understand and predict specificity in protein–DNA interactions, computational approaches can be used to estimate the differences in free energy associated with changes in the DNA-binding site. Starting with the structure of a complex, its free energy is estimated, for example, using MC methods, and then the change in free energy is modeled when single or multiple changes are made to the DNA sequence, which may require conformational changes in the complex. If the overall conformation remains largely unchanged, this approach can often give good estimates of specificity, but if significant rearrangements occur, the prediction accuracy deteriorates. Similar analyses can be used to study the effects of changing specific amino acids in the DNA-binding domain of the protein and to design proteins with altered specificity. There are some notable successes for this approach, such as the design of an endonuclease to recognize and cut DNA at a novel target sequence with similar specificity of the wild-type protein for its natural target (Ashworth et al. 2006). But in general this is a very challenging problem, and the prediction accuracy based on molecular modeling simulations is generally not as high as those obtained by the statistical approaches described in Chapter 7. There are significant advantages in applying modeling simulations to this problem, including the generalization of such methods to a wide range of proteins without the need for extensive experimental data. Continuing research in this area will likely generate significant improvements in the coming years.

Kinetics of Binding-Site Location

For the regulatory system in a cell to function properly, it must not only solve the thermodynamic constraint of having the required binding sites occupied at the appropriate level, but it must also solve the kinetic constraint of occupying the appropriate sites rapidly. In bacteria, a newly synthesized TF or an existing one that is activated for DNA binding can locate its proper binding site, amidst the millions of other competing DNA sites, within a few seconds. Similar rates, up to perhaps a few minutes, are also observed in eukaryotic nuclei that may contain hundreds to thousands of times more genomic DNA. The thermodynamic constraint is solved by making the free energy of binding sufficiently low at the required binding sites to overcome the enormous excess of nonfunctional genomic sites (as well as by synthesizing more than the minimum amount of TF needed to occupy those sites; there is always some TF, usually the majority, that is bound at nonfunctional sites). The low free energy of binding to the functional sites is accomplished using the variety of mechanisms, including contributions from both enthalpy and entropy, described in the preceding sections.

The kinetic constraint is made more difficult because of the thermodynamic constraint. If the only requirement was that the TF encounter all of the sites in the genome, this could be accomplished very rapidly by simple diffusion. Early in vitro experiments with the Lac repressor showed that the protein could apparently locate its binding site faster than allowed by a simple diffusion process. This led to the demonstration that the protein could bind to the DNA at nonoperator positions and transfer along the DNA until it reached the binding site, making the process one-dimensional (1D) diffusion along the DNA molecule rather than three-dimensional (3D) diffusion through the solution. However, the fast diffusion result was obtained under very low salt conditions; at higher salt conditions (close to physiological conditions), the rate of binding-site location was close to that expected from 3D diffusion. As described above, Lac repressor binding affinity is very sensitive to the salt concentration, especially nonspecific binding, so at low salt conditions it will likely stay associated with DNA longer, allowing for increased 1D diffusion. There is also an electrostatic attraction in solution that draws the protein to the DNA faster than random 3D diffusion, and both contribute to the fast association rate measured (Halford 2009).

Theoretical calculations show that 1D diffusion on chromosome-size fragments would hinder, not facilitate, binding-site location. Because 1D diffusion is a random walk, the distance covered along the DNA in N steps would be $\sqrt{N}$ on average, so longer DNA would impede finding the binding sites. Theoretical analyses have also shown that an optimal strategy for locating sites quickly involves a combination of 1D and 3D diffusion, with the amount of time spent in each mode being about equal. Calculations based on this result suggest that the 1D search should be limited to about 100 bp before leaving the DNA, doing a 3D diffusion, and finding another region in which to perform 1D search (Fig. 6-3). Experimental data are consistent with that analysis, suggesting that TFs have evolved to solve the search problem close to optimally (Tafvizi et al. 2011). It is interesting that in eukaryotic genomes, which are largely complexed with histones and inaccessible for TF binding, the nucleosome-free segments constituting the regulatory regions are typically about that size. This suggests

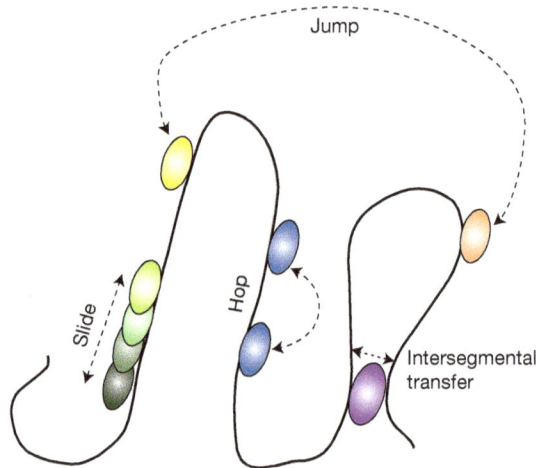

Figure 6-3. Proteins can move from one site on DNA to other sites by a variety of mechanisms. They can "slide" along the DNA, remaining bound nonspecifically while scanning the sequence for good binding sites; "hop" over short distances in which they dissociate from the DNA but remain in close proximity and rebind a short distance away; or "jump" over long distances, essentially a 3D diffusion until a new segment of DNA is encountered and bound. This last mechanism may be between distant segments of the same chromosome or between chromosomes. Some proteins can bind to more than one segment of DNA at a time and therefore may move from one segment to a distant segment without dissociating from the DNA via intersegmental transfer. (Reproduced, with permission, from Tafvizi et al. 2011 [©2011 Wiley-VCH Verlag GmbH & Co. KGaA, Weinheim].)

that a TF might do 3D diffusion between accessible regions and then 1D scanning for optimal binding sites within those regions until it locates its functional site.

Although the 1D/3D search mechanism allows the TF to scan the genome efficiently, the thermodynamic constraint requiring low-energy binding at the functional sites works to oppose the kinetic constraint. The main problem is that there is a limit to the degree of specificity that can be obtained by a DNA-binding protein. Restriction enzymes have the largest degree of specificity for any DNA-binding proteins, with even single base changes to the preferred site having about 10^6-fold reduced activity (that is, the level of reduced cutting of the double-stranded DNA; nicking of one strand is more common, and simply binding is not as specific as cutting). This is accomplished by every base pair of the recognition site making multiple specific contacts with the protein (see Fig. 4-1). But TFs do not achieve such dramatic specificity. In general functional sites are collections of different sequences that may differ from one another at one or more positions (covered in more detail in Chapters 5 and 7). This has the advantage for cells that functional sites are "programmable," which means that different sites have different affinities and therefore respond (or more correctly their occupancy varies) at different concentrations of TF (Gerland et al. 2002). At the low end of functional concentration, some may be bound and others free, whereas at higher concentrations they may all be bound at similar (close to saturating) levels. This allows the cell to turn on (or off) genes in a specific order, or to different total extents. But this raises a large problem in terms of kinetics. To find the functional sites through 1D scanning, the TF must pass over many intervening sites, some of which may be not too different from the correct sites. This has the potential of being a kinetic block, essentially a trap in which the TF will spend considerable time before escaping to continue the search. The rate of escape is related to the binding affinity, the dissociation rate, which is in turn related to the K_D (which is the ratio of dissociation to association rates). If the association rates are diffusion controlled, then differences in K_D are proportional to differences in rates. To keep the search fast, the free energy differences between the sites should be on the order of $k_B T$ (k_B is the Boltzmann constant, equivalent to the gas constant R [see Box 6-1], but on a per molecule basis instead of per mole; in typical biological units this is about 0.62 kcal/mol at body temperature, $310°K$). But this is in opposition to the other constraint of maximizing the free energy difference between the functional sites and the background, which should be at least $5-6k_B T$ to achieve high occupancy of the functional sites. This conflict in constraints gives rise to the need for alternative modes of DNA binding, in which the TF moves quickly over large sequence regions while in a nonspecific binding mode, but must also convert quickly to sequence-specific mode to "read" the sequence and determine if it is a suitable place to stop for an extended period. The switch between the modes must be fast and stochastic, although perhaps some features of the functional sites are visible in the nonspecific mode, such as the shape of the backbone, which can induce switching to the reading confirmation (Gerland et al. 2002; Tafvizi et al. 2011).

REFERENCES

Ashworth J, Havranek JJ, Duarte CM, Sussman D, Monnat RJ Jr, Stoddard BL, Baker D. 2006. Computational redesign of endonuclease DNA binding and cleavage specificity. *Nature* **441:** 656–659.

Bergqvist S, O'Brien R, Ladbury JE. 2001. Site-specific cation binding mediates TATA binding protein-DNA interaction from a hyperthermophilic archaeon. *Biochemistry* **40:** 2419–2425.

Deleeuw L, Tchernatynskaia AV, Lane AN. 2008. Thermodynamics and specificity of the Mbp1-DNA interaction. *Biochemistry* **47:** 6378–6385.

Dragan AI, Frank L, Liu Y, Makeyeva EN, Crane-Robinson C, Privalov PL. 2004a. Thermodynamic signature of GCN4-bZIP binding to DNA indicates the role of water in discriminating between the AP-1 and ATF/CREB sites. *J Mol Biol* **343:** 865–878.

Dragan AI, Read CM, Makeyeva EN, Milgotina EI, Churchill ME, Crane-Robinson C, Privalov PL. 2004b. DNA binding and bending by HMG boxes: Energetic determinants of specificity. *J Mol Biol* **343:** 371–393.

Dragan AI, Li Z, Makeyeva EN, Milgotina EI, Liu Y, Crane-Robinson C, Privalov PL. 2006. Forces driving the binding of homeodomains to DNA. *Biochemistry* **45:** 141–151.

Feig M, Brooks CL 3rd. 2004. Recent advances in the development and application of implicit solvent models in biomolecule simulations. *Curr Opin Struct Biol* **14:** 217–224.

Fujii S, Kono H, Takenaka S, Go N, Sarai A. 2007. Sequence-dependent DNA deformability studied using molecular dynamics simulations. *Nucleic Acids Res* **35:** 6063–6074.

Gerland U, Moroz JD, Hwa T. 2002. Physical constraints and functional characteristics of transcription factor-DNA interaction. *Proc Natl Acad Sci* **99:** 12015–12020.

Halford SE. 2009. An end to 40 years of mistakes in DNA-protein association kinetics? *Biochem Soc Trans* **37:** 343–348.

Houbaviy HB, Burley SK. 2001. Thermodynamic analysis of the interaction between YY1 and the AAV P5 promoter initiator element. *Chem Biol* **8:** 179–187.

Hyre DE, Spicer LD. 1995. Thermodynamic evaluation of binding interactions in the methionine repressor system of *Escherichia coli* using isothermal titration calorimetry. *Biochemistry* **34:** 3212–3221.

Jayaram B, Jain T. 2004. The role of water in protein-DNA recognition. *Annu Rev Biophys Biomol Struct* **33:** 343–361.

Jen-Jacobson L, Engler LE, Ames JT, Kurpiewski MR, Grigorescu A. 2000. Thermodynamic parameters of specific and nonspecific protein-DNA binding. *Supramol Chem* **12**: 143–160.

Kalodimos CG, Boelens R, Kaptein R. 2004. Toward an integrated model of protein-DNA recognition as inferred from NMR studies on the *Lac* repressor system. *Chem. Rev.* **104:** 3567–3586.

Kunne AG, Sieber M, Meierhans D, Allemann RK. 1998. Thermodynamics of the DNA binding reaction of transcription factor MASH-1. *Biochemistry* **37:** 4217–4223.

Ladbury JE, Wright JG, Sturtevant JM, Sigler PB. 1994. A thermodynamic study of the trp repressor-operator interaction. *J Mol Biol* **238:** 669–681.

Liggins JR, Privalov PL. 2000. Energetics of the specific binding interaction of the first three zinc fingers of the transcription factor TFIIIA with its cognate DNA sequence. *Proteins* **Suppl 4:** 50–62.

Merabet E, Ackers GK. 1995. Calorimetric analysis of λ cI repressor binding to DNA operator sites. *Biochemistry* **34:** 8554–8563.

Oda M, Furukawa K, Ogata K, Sarai A, Nakamura H. 1998. Thermodynamics of specific and nonspecific DNA binding by the c-Myb DNA-binding domain. *J Mol Biol* **276:** 571–590.

Ponder JW, Case DA. 2003. Force fields for protein simulations. *Adv Protein Chem* **66:** 27–85.

Privalov PL, Jelesarov I, Read CM, Dragan AI, Crane-Robinson C. 1999. The energetics of HMG box interactions with DNA: Thermodynamics of the DNA binding of the HMG box from mouse sox-5. *J Mol Biol* **294:** 997–1013.

Privalov PL, Dragan AI, Crane-Robinson C, Breslauer KJ, Remeta DP, Minetti CA. 2007. What drives proteins into the major or minor grooves of DNA? *J Mol Biol* **365:** 1–9.

Rohs R, West SM, Liu P, Honig B. 2009. Nuance in the double-helix and its role in protein–DNA recognition. *Curr Opin Struct Biol* **19:** 171–177.

Rohs R, Jin X, West SM, Joshi R, Honig B, Mann RS. 2010. Origins of specificity in protein-DNA recognition. *Annu Rev Biochem* **79:** 233–269.

Seldeen KL, McDonald CB, Deegan BJ, Farooq A. 2009. Single nucleotide variants of the TGACTCA motif modulate energetics and orientation of binding of the Jun-Fos heterodimeric transcription factor. *Biochemistry* **48:** 1975–1983.

Shi Y, Wang S, Krueger S, Schwarz FP. 1999. Effect of mutations at the monomer-monomer interface of cAMP receptor protein on specific DNA binding. *J Biol Chem* **274:** 6946–6956.

Tafvizi A, Mirny LA, van Oijen AM. 2011. Dancing on DNA: Kinetic aspects of search processes on DNA. *Chem Phys Chem* **12:** 1481–1489.

Velázquez-Campoy A, Ohtaka H, Nezami A, Muzammil S, Freire E. 2004. Isothermal titration calo-rimetry. *Curr Protoc Cell Biol*, pp. 17.8.1–17.8.24. John Wiley & Sons, Inc., New York.

Weiss MA, Ellenberger T, Wobbe CR, Lee JP, Harrison SC, Struhl K. 1990. Folding transition in the DNA-binding domain of GCN4 on specific binding to DNA. *Nature* **347**: 575–578.

ONLINE RESOURCES

ProNIT is a database of experimental parameters from published studies of protein–DNA interac-tions. It is available at http://gibk26.bio.kyutech.ac.jp/jouhou/pronit/pronit.html. The following article includes a description of ProNIT:

Kumar MD, Bava KA, Gromiha MM, Prabakaran P, Kitajima K, Uedaira H, Sarai A. 2006. Pro-Therm and ProNIT: Thermodynamic databases for proteins and protein-nucleic acid interactions. *Nucleic Acids Res* **34**: D204–D206.

FURTHER READING

The following articles and reviews, although not specifically cited, contributed to the information provided in this chapter and offer expanded descriptions and discussions on many of the topics.

Bintu L, Buchler NE, Garcia HG, Gerland U, Hwa T, Kondev J, Phillips R. 2005. Transcriptional reg-ulation by the numbers: Models. *Curr Opin Genet Dev* **15**: 116–124.

Bintu L, Buchler NE, Garcia HG, Gerland U, Hwa T, Kondev J, Kuhlman T, Phillips R. 2005. Transcrip-tional regulation by the numbers: Applications. *Curr Opin Genet Dev* **15**: 25–35.

Jelesarov I, Bosshard HR. 1999. Isothermal titration calorimetry and differential scanning calorimetry as complementary tools to investigate the energetics of biomolecular recognition. *J Mol Recognit* **12**: 3–18.

Jen-Jacobson L. 1999. Protein-DNA recognition complexes: Conservation of structure and binding energy in the transition state. *Biopolymers* **44**: 153–180.

Jen-Jacobson L, Engler LE, Jacobson LA. 2000. Structural and thermodynamic strategies for site-specific DNA binding proteins. *Structure* **8**: 1015–1023.

Oda M, Nakamura H. 2000. Thermodynamic and kinetic analyses for understanding sequence-specific DNA recognition. *Genes Cells* **5**: 319–326.

Privalov PL. 2009. Microcalorimetry of proteins and their complexes. *Methods Mol Biol* **490**: 1–39.

Privalov PL, Dragan AI, Crane-Robinson C. 2011. Interpreting protein/DNA interactions: Distinguish-ing specific from non-specific and electrostatic from non-electrostatic components. *Nucleic Acids Res* **39**: 2483–2491.

Sarai A, Kono H. 2005. Protein-DNA recognition patterns and predictions. *Annu Rev Biophys Biomol Struct* **34**: 379–398.

Spolar RS, Record MT Jr. 1994. Coupling of local folding to site-specific binding of proteins to DNA. *Science* **263**: 777–784.

CHAPTER 7

Bioinformatics of DNA-Binding Sites

THIS CHAPTER BEGINS THE THIRD TYPE OF ANALYSIS OF DNA–protein interactions: one using bioinformatics approaches. Although definitions of bioinformatics vary, in the field of protein–DNA interactions this term usually refers to statistical approaches to modeling and predicting binding sites based on large data sets. The modeling may be based on sets of known sites or require motif discovery, in which regions containing binding sites are known but actual sites must be inferred. The data may be only qualitative or collections of sites, or there may also be quantitative information such as binding affinities or probabilities using methods described in Chapter 5. The main goal of these approaches is to facilitate the prediction of binding sites in cases where experimental data are lacking or incomplete. Combined with information about the structural interactions between protein families and their binding sites, as described in Chapters 2 and 3, bioinformatics approaches can also lead to the prediction of specificity based on protein sequences and to the design of proteins with desired specificity, as described in Chapter 8. Good specificity models can also predict the effects of DNA mutations on binding activity and be useful for the design of regulatory systems, topics described in more detail in Chapter 9.

REPRESENTING THE SPECIFICITY OF TRANSCRIPTION FACTORS

The specificity of a transcription factor can be represented in various forms, from a simple consensus sequence to a list of the affinities for all possible binding-site sequences. The complexity of the representation (the number of parameters that the representation contains) varies accordingly, whereas the accuracy will depend on the particular protein. Using restriction enzymes as an example, a simple consensus sequence represents the specificity of cutting very well (binding is a bit more

Table 7-1. Degeneracy code for nucleotides

Code letter	Bases covered	Rationale
A	A	
C	C	
G	G	
T (U)	T (U)	T for DNA, U for RNA
R	A,G	puRine
Y	C,T	pYrimidine
W	A,T	Weak
S	C,G	Strong
M	A,C	aMino group
K	G,T	Keto group
B	C,G,T	not A
D	A,G,T	not C
H	A,C,T	not G
V	A,C,G	not T (or U)
N	A,C,G,T	aNy base

complex). EcoRI cuts at the site GAATTC, and because this specificity is essentially all or none, that is an accurate representation. Even for enzymes with degenerate sites, such as HincII, which cuts at GTYRAC (see Table 7-1 for the degeneracy code), that is an accurate representation because it is still all or none; each of the four sites represented by the consensus are cut equally well, and essentially no other sites are cut (the error rate is very small under normal conditions). In computer science, such patterns (character strings) are called regular expressions, which can even allow for gaps (consensus sequences with variable spacing between different regions, such as TGTGA N[6-7] TCACA, which allows either six or seven bases between the two half-sites). Very fast search algorithms exist that find all matches to a regular expression in any sequence, and a mismatch tolerance can be included. However, the accuracy of the predictions suffers if the consensus sequence is not a good representation of the specificity. As pointed out previously, transcription factors tend not to have as high a specificity as restriction enzymes. One can still define a consensus sequence (either a simple DNA sequence or one allowing degeneracy) that captures the highest affinity sites, but in general there is a much more gradual decrease in affinity as variations are made to the consensus than in the all-or-none activity of restriction enzymes.

POSITION WEIGHT MATRIX

A position weight matrix (PWM) (also known as simply a weight matrix, a position-specific scoring matrix [PSSM], or a position-specific weight matrix [PSWM]) is a step more complex than a simple consensus sequence (Fig. 7-1A) because it gives a score for all possible bases at every position in the binding site. Figure 7-1B shows the use of the PWM to scan along a sequence and give a score to each position,

A

Base/position	1	2	3	4	5
A	−1.7	−2.7	−1.7	−0.7	1.5
C	−1.1	−1.7	−1.7	−2.7	−1.7
G	−2.7	1.5	−2.7	1.5	−1.7
T	1.6	−0.4	1.7	−1.7	−1.1

B

.... C **A** **T** **G** **T** **G** A C .... = −8.2

.... C **A** **T** **G** **T** **G** **A** C ... = 7.8

.... C A **T** **G** **T** **G** **A** **C** .... = −8.2

C $S_i = \text{TGTGA}$

Base/position	1	2	3	4	5
A	0	0	0	0	1
C	0	0	0	0	0
G	0	1	0	1	0
T	1	0	1	0	0

$$\text{Score}(S_i \mid W) = \vec{S_i} \cdot \vec{W} = 7.8$$

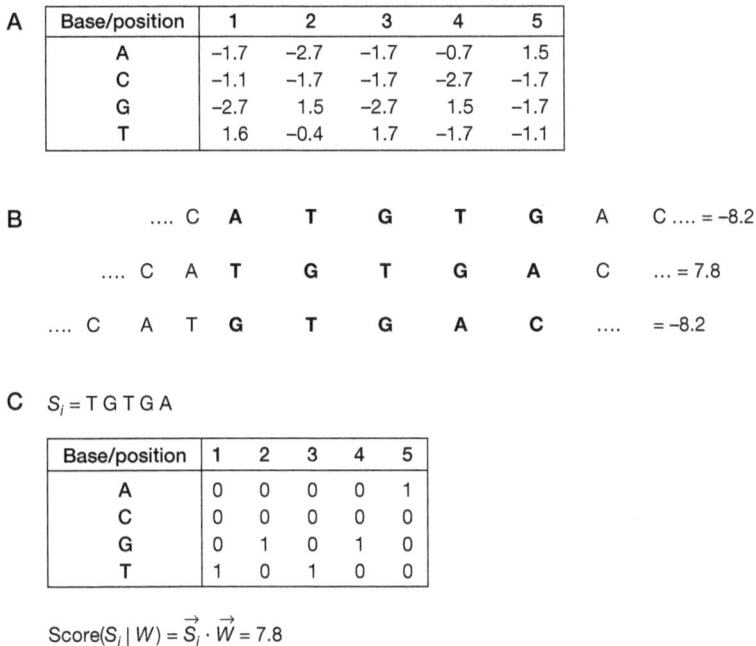

Figure 7-1. Position weight matrix (PWM) model of transcription-factor binding specificity (A) PWM for a protein that contains a score for each possible base (A, C, G, or T) at each of the five positions of the binding site. (B) A segment of the genome is scored by the matrix in each of three consecutive positions. The score for each 5-long subsequence (boldface) aligned with the PWM is obtained by summing the elements of the PWM that correspond to the base occurring at each aligned position. (C) Encoding the specific sequence TGTGA as a matrix with the same rows and columns as the PWM. The base that occurs at each position is assigned a value of 1 and all other bases at each position are assigned a value of 0. The score of the sequence, given the matrix, is the dot product between the matrices (the sums of the products of the corresponding elements).

including the highest possible score, 7.8, for the consensus sequence of the PWM, TGTGA. Labeling sites as binding or not binding requires a threshold, so that sites with scores above the threshold would be considered binding and those below would be considered not binding. With such a dichotomy, any consensus representation, including degeneracies (and even gapped patterns with a modification to the PWM), can be represented exactly with an equivalent PWM. Therefore, there is no advantage in accuracy by using a consensus sequence instead of a PWM. The advantage of a PWM is that its scores can be different for every sequence and a threshold need not even be defined because the scores, if the PWM represents the specificity well, could be directly related to binding affinity or binding probability. Predicting binding affinity and binding probability are related, but somewhat different, goals. Binding probability will depend on the concentration of the transcription factor, whereas binding affinity does not (see Chapter 5). At very low concentrations of the protein, the binding probability and binding affinity will be proportional, but at high concentrations they can be very different (see Eq. 5-1 and Fig. 5-1). Knowing the affinity allows one to predict the binding probability for a sequence at any given protein concentration; thus, it is the more fundamental measure. But PWMs based on probabilities are commonly used and can be useful models of specificity, at least for some proteins and under some conditions.

Any sequence can be encoded in the same matrix form as the PWM by having a single 1 in each position corresponding to the base that occurs there, with all other elements being 0 (Fig. 7-1C). This can obviously be generalized to a degenerate sequence by allowing more than one base in a column to be 1 (or other nonzero value, perhaps summing to 1). We can then consider both the sequence and the PWM to be vectors (e.g., by concatenating the columns) such that the score given to a sequence S_i, by the PWM W, is simply $\vec{W} \cdot \vec{S}_i$, the dot product between the two vectors. This is a useful way of thinking about the use of PWMs because there are many algorithms dealing with vectors that can be applied. It is also easy to see how the idea of a PWM can be generalized to a weight vector that can score a sequence based on whatever features are important for the function being considered. In the standard PWM, the features are simply which base occurs at each position in the site, but the sequence could be encoded to capture other types of features, such as combinations of bases, the composition of segments, the base-pairing potential (complementarity) of specific positions, or any other characteristics of a sequence. If the features can be defined such that they contribute additively to the functional activity of the sequence, a vector can be determined that weights each feature appropriately so that the sum of the weighted features, the dot product of the feature vector and the weight vector, accurately predicts the functional activity of each sequence.

There are a variety of computational approaches, including some machine learning algorithms and other statistical methods, that can take input collections of sequences and their functional activities and determine the important features and how to weight them to make accurate predictions. In the next few sections of this chapter, we assume that the important features are simply the base that occurs at each position, so that standard PWMs can capture the functional activity, meaning that the total activity is just the sum of the contribution from each base in the sequence. But binding activities do not have to be that simple, and later in this chapter we consider some more complex models for protein–DNA interactions.

Keep in mind is that a simple PWM, with four values at each position, is not unique. For example, if a constant is added to every element of position 1 and the same constant is subtracted from every element of position 2, the scores of all sequences would remain the same. One can add constraints to make it unique, and there are various ways that this is done. For example, in energy modeling, it is common to assign an energy value of 0 to the preferred base at each position, and then the other values in the PWM correspond to the difference in binding energy between the preferred base and each of the other bases. In a probabilistic model (see below), in which the PWM elements correspond to the log(probability) of each base at each position, the exponentiated elements at each position will sum to 1. A log-odds PWM has a similar constraint except that the background probabilities are included before exponentiation.

Modeling Specificity with a PWM

Given that a PWM will be used to model the specificity of a transcription factor, the main question is how to obtain the elements of the PWM. There have been many methods developed over the years, but the best one depends on the type of data available and perhaps the purpose for which the PWM will be used. The following sections present some general classes of methods.

Probabilistic Models

One of the most commonly used approaches is to model the specificity probabilistically based on a collection of known binding sites. In this approach, we assume that the collection of known sites is just a sample and that there are other sites that we have not yet observed. We want to use this sample to estimate the probability that any other site would be a binding site based on its similarity to the sample. As an example, Figure 7-2A shows a (partial) list of sequences that are binding sites for some transcription factor, Figure 7-2B shows the "count matrix" (the number of occurrences of each base at each position in the full set of 22 known binding sites), and Figure 7-2C converts this to the probabilities, often referred to as a position frequency matrix (PFM). If the positions contribute independently to the activity (as we are currently assuming), the probability of seeing a particular base b at position j of a binding site would be the element $F(b,j)$. And the probability of sequence S_i belonging to the collection of functional sites would be the product of the probabilities for each base at each position in S_i. This could be represented as

$$\Pr(S_i) = \prod_{b,j} \vec{F}^{\vec{S_i}},$$ (7-1)

where $\vec{F}$ is the vector encoding the PFM. The calculation of the probability that S_i is a binding site takes the product of the elements of $\vec{F}$ that correspond to the sequence vector $\vec{S_i}$. This can be converted to the standard PWM approach if we set $W_{LP}(b,j) = \log F(b,j)$, with the resulting score being the log(probability) that the sequence is in the functional set.

$$\text{Score}(S_i|W_{LP}) = \vec{W}_{LP} \cdot \vec{S_i} = \log \Pr(S_i),$$ (7-2)

where LP indicates that W is based on log probabilities. PWMs generated by other methods have alternative subscripts to indicate the method.

A

```
T G T G G
T T T G A
A G T G T
T T T G C
T G T G A
T G C A A
T G T T A
T T T G A
T G T G A
    ⋮
    ⋮
    ⋮
```

B   Count matrix $N(b,j)$

Base/position	1	2	3	4	5
A	1	0	1	3	18
C	2	1	1	0	1
G	0	17	0	18	1
T	19	4	20	1	2

C   Frequency matrix $F(b,j)$

Base/position	1	2	3	4	5
A	0.05	0	0.05	0.14	0.81
C	0.09	0.05	0.05	0	0.05
G	0	0.77	0	0.81	0.05
T	0.86	0.18	0.90	0.05	0.09

D   Log-odds matrix $W_{LO}(b,j) = \log F'(b,j)/F(b,0)$

Base/position	1	2	3	4	5
A	−1.7	−2.7	−1.7	−0.7	1.5
C	−1.1	−1.7	−1.7	−2.7	−1.7
G	−2.7	1.5	−2.7	1.5	−1.7
T	1.6	−0.4	1.7	−1.7	−1.1

Figure 7-2. Making PWMs from known sites. (A) Partial list of binding sites for a hypothetical TF. (B) The count matrix for the complete list of all 22 known sites. The elements of the matrix are the number of occurrences of each base at each position in the alignment. (C) The frequency matrix is the count matrix divided by the total number of sequences (here, rounded to two significant digits). (D) Log-odds matrix based on the frequency matrix of C. To avoid taking the logarithm of 0, the counts are all augmented by +1 and the total number of sites is increased to 26. (Put another way, a pseudocount of +1 is added to each of the real counts for each base at each position, which increases the total counts at each position to 26.) The elements of the PWM are the logarithms of the ratio between the observed frequencies (including pseudocounts and labeled as F′) and the expected frequencies based on the genomic, or background, frequencies for each base. In this case, the background is assumed to be 0.25 for each base.

This may seem to be a bit strange. We start with a collection of known binding sites and then build a model that assigns probabilities to sequences being binding sites, but no sequence would get a probability of 1 (except in the case of completely conserved sites, such as the EcoRI sites described above), even those that we knew were binding sites. But this type of modeling makes sense when we consider binding sites from two different viewpoints. One viewpoint is quantitative binding affinity, as described in Chapter 5. Equation 5-4 described the probability of a particular sequence being bound when all possible sites are competing for binding to a single protein. In that case, the sum of the binding probabilities of all sequences is equal to 1, which is also the case for Equation 7-1. Thus, in both cases, the calculated probabilities can be used to rank different sequences based on their probabilities. The true probability of a sequence being bound, either in vitro or inside a cell, depends on the concentration of the protein, the set of other potential binding sites that are competing for the same pool of proteins, and any other proteins that may influence its binding through competition or cooperativity. In Chapter 5, the probability was based on knowing the affinities of all possible binding sites, whereas in Equation 7-1 we assume that the set of known binding sites, from which the PFM is derived, is representative of high-affinity sites from which we can estimate the affinity contributions of each base at each position. If the new sequence has a probability, from the PFM, that is in the same range as the known binding sites, it is also likely to be a binding site. This sets a useful threshold.

The other relevant viewpoint for probabilistic modeling is purely statistical and has no reliance on binding affinities. This approach is called a generative model where one imagines a machine that generates binding sites and does so based on the probabilities of bases occurring at each position of the site. Such a machine can generate any sequence (if there are no zero probability bases at any of the positions), but some will have much higher probabilities than others. Again, those sites with probabilities within or near to the range of the known sites are also likely to be binding sites. When considering binding sites contained within a genome, it is also useful to consider the genomic, or background, probability of bases. If $F(b,j)$ is the frequency of each base b at each position j in the set of known binding sites, we call $F(b,0)$ the frequency of each base (the composition) in the genome as a whole (or really, the rest of the genome minus the binding sites, but because the sites are a very small fraction of the total genome, they make almost no contribution to the genomic frequencies). We can imagine another machine that generates this background sequence, and then the probability can be calculated for a particular site being generated by both machines. The ratio of those probabilities of a sequence coming from the site-generating machine versus those coming from the genome-generating machine provides a very useful measure for the specificity of the protein and for the significance of the set of binding sites used to generate the PFM (see Box 7-1). When performing motif discovery (see below), such a significance measure is critical to identifying the most likely binding sites. Rather than computing the site and genome probabilities inde-

---

Box 7-1. Likelihood Ratios and Information Content

Assume that we have a data set with D total observations and D_i observations with outcome i. For example, if we are flipping a coin, i could be either heads or tails. D_{heads} is the number of observed heads and D_{tails} the number of observed tails; $D = D_{heads} + D_{tails}$. D might also be the result of rolling a die and i would be the set of possible outcomes, or any integer from 1 to 6.

Let P_i be a probability distribution over the set of possible outcomes: $\sum_i P_i = 1$. We can calculate the probability of observing any particular data set D:

$$\Pr\left(D|P_i = \frac{D!}{\prod_i D_i!} \prod_i P_i^{D_i}\right),$$

where $\Pr(D|P_i)$ is the probability of observing the data set D (defined by "|") given the probability distribution P_i. The first term is the number of different orders of observations that give the same number of observations of each type. For example, if the data are three heads and two tails in five flips of a coin, there are 10 different orders of heads and tails that could give the same set of observations ($5!/3!2!=10$). If we want to compare two different probability distributions, say, P_i and Q_i, to determine which is more likely to have generated the data, we can compute the ratio of their probabilities. In this case, the first term cancels out because it depends only on the data, which is the same in the two probability calculations, and is independent of P_i and Q_i. The likelihood ratio (LR) of observing the data D for the two different probability distributions is

$$LR(D|Q_i, P_i) = \frac{\Pr(D|Q_i)|}{\Pr(D|P_i)|} = \prod_i \left(\frac{Q_i}{P_i}\right)^{D_i}.$$

It is often more convenient to compute the log-likelihood ratio (LLR):

$$LLR(D|Q_i, P_i) = \ln\left(\frac{\Pr(D|Q_i)}{\Pr(D|P_i)}\right) = \sum_i D_i \ln\left(\frac{Q_i}{P_i}\right).$$

This is true for any two probability distributions, so it can be used to identify the most likely probability distribution for a given data set D. It is easy to show (see the note at the end of the box) that the most probable distribution, called Q_i^*, is $Q_i^* = D_i/D$. Therefore, the LLR between the most likely probability distribution and any other distribution P_i is

$$LLR(D|Q_i^*, P_i) = D\sum_i Q_i^* \ln\left(\frac{Q_i^*}{P_i}\right) \geq 0.$$

If P_i is a null or background distribution, $2 \cdot LLR(D|Q_i^*, P_i)$ is the G statistic, closely related to χ^2, that is used to determine the statistical significance of the data set D given an expected data set based on the background distribution. $LLR(D|Q_i,P_i)/D$ (just the formula inside the sum, leaving out the sample size D) is also known as the Kullback–Leibler distance, $KL(Q\|P)$, between the two probability distributions (sometimes referred to as KL divergence; some object to calling it a distance because it is not symmetric). In computer science and information theory, it is also referred to as relative entropy.

Relative entropy, or KL(Q‖P), is exactly the same as information content as defined in Equation 7-5 below. If multiplied by the sample size, it is the LLR($D|Q_i,P_i$) described above and a very useful statistic for judging the statistical significance of the binding-site position probabilities compared to what would be expected from choosing sites at random from the genome. When performing motif discovery, it is a valuable criterion for determining which of the possible binding-site alignments is most significant.

Note: $Q_i^* = D_i/D$ is the most probable distribution:

$$x - 1 \geq \ln x \qquad \forall x > 0.$$

Let Q_i and P_i be two probability distributions over the set of possible outcomes i:

$$\sum_i P_i = \sum_i Q = 1.$$

Substitute P_i/Q_i for x in Equation B1:

$$\frac{P_i}{Q_i} - 1 \geq \ln \frac{P_i}{Q_i} \qquad \forall \frac{P_i}{Q_i} > 0.$$

Therefore,

$$\sum_i Q_i \left(\frac{P_i}{Q_i} - 1 \right) = \sum_i (P_i - Q_i) = 0 \geq \sum_i Q_i \ln \frac{P_i}{Q_i},$$

which proves that

$$\sum_i Q_i \ln \frac{Q_i}{P_i} \geq 0$$

and shows that the highest probability for LLR $= \sum_i D_i \ln (Q_i/P_i)$ is when $Q_i = D_i/D$.

pendently, they can be incorporated together into a PWM, referred to as the log-odds PWM (see Fig. 7-2D):

$$W(b,j) = \log[F(b,j)/F(b,0)]. \tag{7-3}$$

This method assures that, at a particular position j, bases that occur at the same frequency in the binding sites as in the genome contribute 0 to the total score, bases that are more frequent in the sites than in the genome contribute a positive score, and bases that occur less frequently in the sites contribute a negative score. The average score over all of the possible genome sites will be negative, whereas the average score of the known sites, those that generated $F(b,j)$, will be positive (see Box 7-1):

$$\text{mean}\left(\vec{W}_{\text{LO}} \cdot \vec{S}^+ \right) = \vec{W}_{\text{LO}} \cdot \vec{F} \geq 0. \tag{7-4}$$

The probabilistic model lends itself to analysis by information theory. A measure of information content can be defined for each position j and summed over all positions to give the information content of the entire site. In fact, multiple definitions of information content have been published (Schneider et al. 1986), but the one most relevant to analyzing the specificity of protein–DNA interactions is based on W_{LO} and is a LLR measure (see Box 7-1). By this definition, the information content (IC) at position j for a set of sites S^+ is

$$IC_j(S^+) = \sum_{b=A}^{T} F(b,j) \log \frac{F(b,j)}{F(b,0)} = \sum_{b=A}^{T} F(b,j) W_{LO}(b,j) \geq 0 \qquad (7\text{-}5)$$

and the IC for the whole motif (summed across the positions) is

$$IC(S^+) = \sum_{j} IC_j(S^+) = \vec{W}_{LO} \cdot \vec{F} \geq 0. \qquad (7\text{-}6)$$

The logarithm can be to any base, but it is common to use base 2 so that IC is measured in bits. IC analysis leads to a convenient graphical method for displaying the specificity of a protein, the "sequence logo" (Schneider and Stephens 1990). Figure 7-3 shows the logo for the protein analyzed in Figure 7-2. The height of the column at each position is the information content of that position, and the height of each base within each column is proportional to its frequency, with higher-frequency bases on top of lower frequencies. The consensus sequence can be simply read across the top of the logo, but, in addition, one can see those positions that are highly conserved (presumably most critical for function) and those that can be substituted with smaller effects on activity.

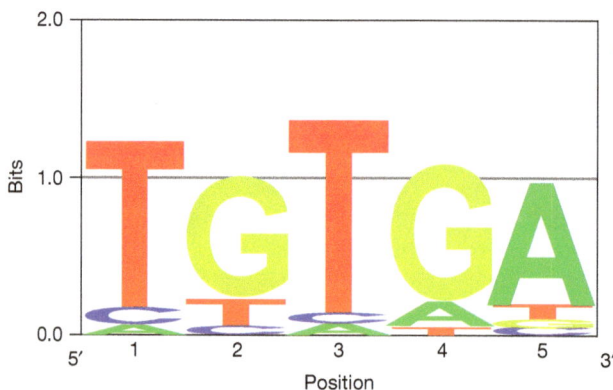

Figure 7-3. Logo representation of the PWM from Figure 7-2D. The height of the stack of bases at each position is the information content of that position, and the heights of the individual bases are in proportion to their probabilities.

This formula for IC is often called relative entropy and is also referred to as the Kullback–Leibler distance between two probability distributions (see Box 7-1). As a LLR statistic, it is used to measure the significance of a binding-site motif, essentially, how different it is from a set of genome sequences chosen at random. It is also related to how often sequences are expected to occur by chance, with their expected frequency being 2^{-IC} (if information content is in bits). This is easily seen for all-or-none cases such as restriction enzymes. For example, the EcoRI site GAATTC has 12 bits of information (if the background is 0.25 for each base) and is expected to occur by chance once every $4096 = 2^{12}$ positions in a random genome. If the genome is AT rich, the IC would be reduced and the expected frequency would be increased. Because information content is the average score of the known binding site, the expected frequency of sites with the average score, or better, is 2^{-IC}. If instead of just a set of binding sites, the PFM is based on all possible sites weighted by their affinity to the protein, IC is equal to the average specific binding energy defined in Chapter 5 (Eq. 5-6) and used as a measure of the specificity of the protein.

Discriminatory Models

If one has sets of sequences that have some function, S^+, and another set lacking that function, S^-, a PWM that separates the two sets is needed. That is, find W_D (for discriminatory W) such that $\vec{W}_D \cdot \vec{S}^+ > \vec{W}_D \cdot \vec{S}^-$ for all S^+ and S^-. This approach was the first use of a PWM, where the elements were the weights of a perceptron, a simple feed-forward neural network designed to learn linear discriminant functions (Stormo et al. 1982). One advantage of this method is that it can easily be extended to include nonadditive models that require nonlinear discriminant functions, by multilayer neural networks (NNs) or another type of machine learning approach called a support vector machine (SVM), and these approaches have been applied in some cases. In fact, the SVM can be used with only positive examples by finding the smallest W_D for which $\vec{W}_D \cdot \vec{S}^+ > 1$ (or any constant). Such a W_D can be obtained by a quadratic programming algorithm, and this is equivalent to training an SVM with only the known (positive) examples (Djordjevic et al. 2003). This method finds the center of the S^+ set, thereby capturing all of the S^+ set while minimizing the total volume of sequence space above the threshold and the number of sites that are predicted to be functional.

Regression Models

If one has not only a collection of binding sites but also for each one a measure of its binding affinity, multiple linear regression methods can be applied to obtain the PWM that provides the best fit to the data. In this case, binding affinities should be converted to binding energies because we expect those to be additive (contributions to affinity are like probabilities and the elements are multiplied together to get the overall affinity, whereas log[affinity] is proportional to binding free energy; see Chapter 5). Given the set of binding energies E_i for each sequence S_i, multiple linear regression will find the energy matrix $E(b,j)$ (the binding energy contribution of each base b and each position j) that we can simply call $\vec{E}$, which minimizes.

$$\sum_i (E_i - \vec{E} \cdot \vec{S}_i)^2, \tag{7-7}$$

the sum of the squared differences between the measured energies and the energies predicted by the model (the energy PWM). If the best energy model only gives a poor fit to the data, that fit implies that the energy is not additive and that no PWM can provide a good fit to the data. In this case, a more complex model must be used to obtain good predictions of binding energies (see below).

Suppose that one has quantitative binding data in the form of binding probabilities, not directly affinities, for a large set of (perhaps all possible) binding sites. For example, in the SELEX experiments described in Chapter 5, the probability of a sequence being bound is related to its affinity. One can apply the log-odds approach to that data as follows:

$$W_{LO}(b, j) = \log F(b, j)/F(b, 0), \tag{7-8}$$

and then $E(b,j) = -W_{LO}(b,j)$ is an estimate of the binding energy. (Note the change in sign, because W assigns higher scores to bases with higher probabilities, which corresponds to lower energies.) However, this runs the risk that contributions to log(probability) may not be additive even if the energies are additive because the relationship between log(probability) and energy is not linear. Recall from Chapter 5 that

$$\Pr(S_i \text{ bound}) = \frac{1}{1 + e^{E_i - \mu}} = \frac{e^{-E_i}}{e^{-E_i} + e^{-\mu}}, \tag{7-9}$$

where μ, the chemical potential, is related to the concentration of the protein. If E_i is large compared to μ, which is the case for low-affinity binding sites or for all sites at low concentrations of the binding protein, the denominator can be considered a constant and the probability is proportional to e^{-E_i}, or log(probability) is directly proportional to binding energy. But if μ is large compared to E_i, at high protein concentrations, the probability approaches 1 (saturation) for high-affinity (low-energy) sites and the effects of independent positions may not be additive in log(probability), although they are in energy. As an example, suppose that for some site we have $E - \mu = -3$, so that the probability of being bound is 0.95. If two different variants at different positions each make $E - \mu = -1$, they both reduce binding probability to 0.73. But the double variant with additive energies, so $E - \mu = 1$ would reduce binding probability to 0.27, about half that expected if the probabilities are independent. In this example, the simple probabilistic approach, using observed frequencies of individual sites, would conclude that the positions do not contribute independently to the binding, whereas the binding energy contributions are independent and additive.

Given probabilities of binding to many (possibly all) sites, nonlinear regression can be performed on the data. Assume there is an additive energy matrix and solve for the parameters, including both the energy matrix and μ, that give a best fit to the data. This approach has been applied to some high-throughput sets of binding data to give good estimates of binding energy matrices (Zhao et al. 2009). It can also be used to identify the cases for which a PWM method does not work because the best PWM fits the data poorly and more complex models must be used.

Higher-Order Models

The main limitation of the PWM model for protein–DNA interactions is the assumption of independent contributions to binding from the separate positions in the binding site. This assumption is not likely to be valid in detail, but for many proteins it appears to be a good approximation. However, for other proteins, it provides a poor representation of the specificity and as a result more complex models are needed. Thinking of the weight matrix concept more generally as providing energies for different features of the sequence that can be summed to get the total energy, higher-order combinations of bases can be added until the specificity is well modeled.

Figure 7-4 provides a graphic form of a higher-order model for Hnf4α, a transcription factor for which simple PWMs did not fit the binding data well (Zhao et al. 2012). Although this diagram has some similarity to the logo of Figure 7-3, the information plotted is different. For each position in the binding site, there is an energy contribution for each base at that position; note that lower energies are plotted on top to conform to the bioinformatic convention of having the preferred bases on top. In addition to those individual base contributions, contributions come from the adjacent dinucleotides at positions 4 and 5 that are not captured by the individual bases; that is, those energies are the residuals after the independent base contributions have been taken into account. The predicted energy of binding to any 8-long sequence can be determined by summing the energy values corresponding to that sequence, including both the individual base and dinucleotide contributions. Of course, there could be other contributions to binding energy, but for that particular data set the single dinucleotide contribution captured the binding affinity well—much better than the simple PWM model. For nearly every protein that has been examined in detail, if a simple PWM is not sufficient, a model that includes energies for adjacent dinucleotides captures almost all of the specificity. A few exceptions are cases in which the protein can bind in alternative modes, and in those cases, a pair of PWMs or dinucleotide models can work quite well.

The number of possible binding base combinations grows exponentially with the length of the binding site: There are 4^L possible sites of length L. High-throughput

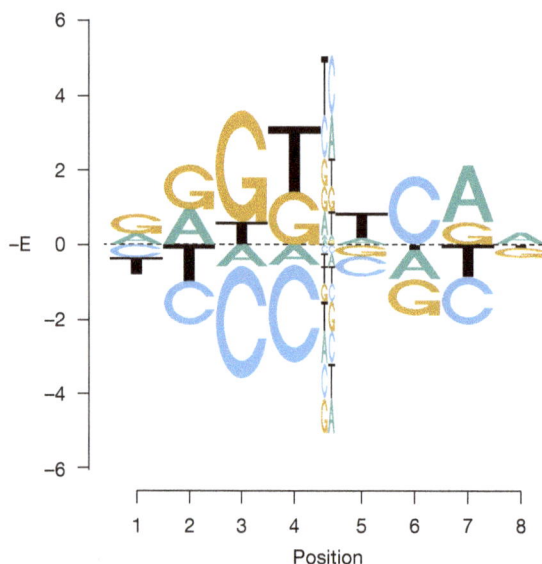

Figure 7-4. Energy logo for Hnf4α (Zhao et al. 2012). The heights of each base are equal to the energy contributions of that base at each position, with the sum of the energies equal to 0. (Note that negative energies, for the preferred bases, are on top.) This logo also includes energies for dinucleotide contributions from adjacent bases at positions 4 and 5. The total predicted energy for any 8-long sequence is the sum of all of the energies corresponding to that sequence, including both mononucleotide and dinucleotide contributions.

binding assays, such as those described in Chapter 5, can collect data on essentially every possible site for proteins with sites up to ~10 long (>10^6 possible sites). With such data, one can ask whether it is even necessary to bother developing a model for the specificity. Why not just take the list of measurements to each site as the specificity of the protein? This is a reasonable idea, but such an approach is often limited by the accuracy of the data. For example, for an 8-long binding site, there are 65,536 different sequences (or approximately half that number when both orientations are considered), and there is a level of uncertainty in each measurement. If the protein did bind additively, so that a PWM was sufficient, each of the 24 (8×3) independent parameters of the PWM would be obtained from averages over thousands of different individual binding sites and could be much more accurate than the individual measurements. Furthermore, if that simple model were adequate, it would provide information about how the protein interacts with the DNA, with the positions being essentially independent of one another. This might also provide for better predictions of binding specificities for proteins with similar sequences that bind to DNA in similar ways (see Chapter 8). Thus, even with enormous amounts of data, it is worthwhile to develop models of the specificity that find the minimum number of parameters that provide an accurate model of the specificity.

DISCOVERY OF BINDING-SITE MOTIFS

In the previous sections, we deal with how to build a model of specificity for a protein when the binding sites are known, and perhaps other data are also known, such as binding affinities or probabilities for many of the binding sites. It is often much easier to identify segments of DNA, such as from in vitro experiments, that contain binding sites, even though the location of the binding sites is not known precisely. These data may also come from a variety of in vivo experiments, such as genetic experiments in which genes regulated by a common factor are identified, expression experiments where genes that appear coregulated are identified, or chromatin immunoprecipitation (ChIP) experiments in which in vivo binding locations of specific transcription factors (TFs) are identified (see Chapter 9).

Figure 7-5A shows an example in which binding sites for the *Escherichia coli* cAMP receptor protein (CRP) are included in a set of upstream regions, but the exact locations are not indicated. This is a case of the "motif discovery" problem (Stormo and Hartzell 1989). If we know where the sites are, we can build a PWM based on them, as in Figure 7-2, or if we know the PWM, we could predict where the binding sites occur in each sequence by simply scoring all of the possible sites and picking the best, as in Figure 7-1B. But given that we do not have either type of information, we need an algorithm that will predict both the positions and the PWM simultaneously. First, we need some objective function, a criterion to evaluate which one of the possible models is best. One good objective function is to find the alignment of sites with the highest IC (see Box 7-1). Figure 7-5B shows the logo of the binding sites for the data set of Figure 7-5A, which was discovered by searching for such an alignment, using one site from each sequence. Other objective functions are possible and may vary depending on the discovery algorithm, but we certainly want to find a model for the sites that is statistically significant, meaning that we are unlikely to find a model that fits the data nearly as well if the sequences were, in fact, randomly chosen instead of being chosen because they contain binding sites. A variety of algorithms have been developed that are described in the sections below. The first general class is based on searching through the set of possible consensus sequences to find the one that is most significant. This can be very effective, but because consensus sequences are often not a good representation of the binding-site motif, one can also search for an alignment of sites, which can be summarized in a PWM, that is most significant. The latter three approaches describe methods that search directly for a PWM that is most significant.

Pattern Searches

If the specificity is to be modeled as a consensus sequence, the most efficient motif discovery method is to search for overrepresented "words," or short sequences that may represent binding sites. The binding-site length for TFs that bind as monomers

A

```
CE1CG
TAATGTTTGTGCTGGTTTTTGTGGCATCGGGCGAGAATAGCGCGTGGTGTGAAAGACTGTTTTTTTGATCGTTTTCACAAAAATGGAAGTCCACAGTCTTGACAG
ECOARABOP
GACAAAAACGCGTAACAAAAGTGTCTATAATCACGGCAGAAAAGTCCACATTGATTATTTGCACGGCGTCACACTTTGCTATGCCATAGCATTTTTATCCATAAG
ECOBGLR1
ACAAATCCCAATAACTTAATTATTGGGATTTGTTATATATAACTTTATAAATTCCTAAAATTACACAAAGTTAATAACTGTGAGCATGGTCATATTTTTATCAAT
ECOCRP
CACAAAGCGAAAGCTATGCTAAAACAGTCAGGATGCTACAGTAATACATTGATGTACTGCATGTATGCAAAGGACGTCACATTACCGTGCAGTACAGTTGATAGC
ECOCYA
ACGGTGCTACACTTGTATGTAGCGCATCTTTCTTTACGGTCAATCAGCAAGGTGTTAAATTGATCACGTTTTAGACCATTTTTTCGTCGTGAAACTAAAAAAACC
ECODEOP2
AGTGAATTATTTGAACCAGATCGCATTACAGTGATGCAAACTTGTAAGTAGATTTCCTTAATTGTGATGTGTATCGAAGTGTGTTGCGGAGTAGATGTTAGAATA
ECOGALE
GCGCATAAAAAACGGCTAAATTCTTGTGTAAACGATTCCACTAATTTATTCCATGTCACACTTTTCGCATCTTTGTTATGCTATGGTTATTTCATACCATAAGCC
ECOILVBPR
GCTCCGGCGGGGTTTTTTGTTATCTGCAATTCAGTACAAAACGTGATCAACCCCTCAATTTTCCCTTTGCTGAAAAATTTTCCATTGTCTCCCCTGTAAAGCTGT
ECOLAC
AACGCAATTAATGTGAGTTAGCTCACTCATTAGGCACCCCAGGCTTTACACTTTATGCTTCCGGCTCGTATGTTGTGTGGAATTGTGAGCGGATAACAATTTCAC
ECOMALBA
ACATTACCGCCAATTCTGTAACAGAGATCACACAAAGCGACGGTGGGGCGTAGGGGCAAGGAGGATGGAAAGAGGTTGCCGTATAAAGAAACTAGAGTCCGTTTA
ECOMALBA
GGAGGAGGCGGGAGGATGAGAACACGGCTTCTGTGAACTAAACCGAGGTCATGTAAGGAATTTCGTGATGTTGCTTGCAAAAATCGTGGCGATTTTATGTGCGCA
ECOMALT
GATCAGCGTCGTTTTAGGTGAGTTGTTAATAAAGATTTGGAATTGTGACACAGTGCAAATTCAGACACATAAAAAAACGTCATCGCTTGCATTAGAAAGGTTTCT
ECOOMPA
GCTGACAAAAAAGATTAAACATACCTTATACAAGACTTTTTTTTTCATATGCCTGACGGAGTTCACACTTGTAAGTTTTCAACTACGTTGTAGACTTTACATCGCC
ECOTNAA
TTTTTTAAACATTAAAATTCTTACGTAATTTATAATCTTTAAAAAAAGCATTTAATATTGCTCCCCGAACGATTGTGATTCGATTCACATTTAAACAATTTCAGA
ECOUXU1
CCCATGAGAGTGAAATTGTTGTGATGTGGTTAACCCAATTAGAATTCGGGATTGACATGTCTTACCAAAAGGTAGAACTTATACGCCATCTCATCCGATGCAAGC
PBR322
CTGGCTTAACTATGCGGCATCAGAGCAGATTGTACTGAGAGTGCACCATATGCGGTGTGAAATACCGCACAGATGCGTAAGGAGAAAATACCGCATCAGGCGCTC
TRN9CAT
CTGTGACGGAAGATCACTTCGCAGAATAAATAAATCCTGGTGTCCCTGTTGATACCGGGAAGCCCTGGGCCAACTTTTGGCGAAAATGAGACGTTGATCGGCACG
TDC
GATTTTTATACTTTAACTTGTTGATATTTAAAGGTATTTAATTGTAATAACGATACTCTGGAAAGTATTGAAAGTTAATTTGTGAGTGGTCGCACATATCCTGTT
```

B

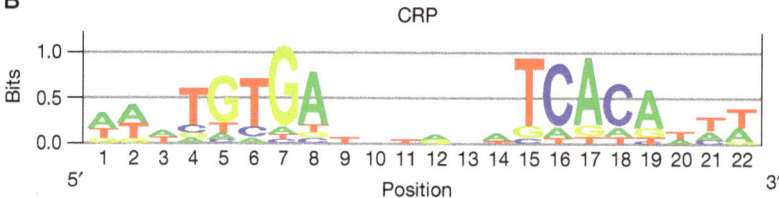

Figure 7-5. Motif discovery problem. (A) Collection of sequences each known to contain a binding site for the CRP protein. In the motif discovery problem, neither the exact locations of the binding sites nor the PWM that represents the CRP motif is known; both are to be inferred from the sequences. (B) Logo of the CRP PWM obtained from the discovered binding sites.

is usually <10 bp, and there are only ~10^6 different 10-long words, so one could imagine searching each sequence with all possible 10-long words and finding the one with the most matches. The problem with that idea is that many (perhaps most) of the binding sites for the TF will not match the consensus sequence exactly and may have some mismatches. In the example of Figure 7-5, the consensus sequence is TGTGA-6N-TCACA, which is a 10-long site, although it contains six non-specific positions in the center (this is because this protein binds as a dimer and each monomer binds to the TGTGA half-site, with the two halves being inverted repeats, the identical pattern on opposite strands). But in the set of sites contained in the sequences of Figure 7-5A, none of them contains an exact match to that consensus.

Several have one mismatch, but even more have two or three mismatches to the consensus. Thus, a strategy to find the consensus sequence by searching through all possible consensus sequences must allow for mismatches. One could count the number of exact matches to each possible consensus, plus the number of single mismatches perhaps weighted by some penalty, the number of double mismatches weighted by a higher penalty, and so on. At the end would be a score for each of the possible consensus sequences based on a combination of its number of occurrences with 0 or more mismatches, with the mismatches being penalized. This is very close to the algorithm RTIDE by Galas, Eggert, and Waterman (Galas et al. 1985) that was shown to be effective in getting both parts of the *E. coli* promoter consensus sequence. The length of the pattern for which to search, the number of mismatches to be allowed, and the penalty for each mismatch are all parameters to be determined and contribute to the statistical significance of the identified pattern.

Of course, as the binding site gets longer, it becomes impractical to search for all possible consensus sequences; at 15-long binding sites, there are $>10^9$ different patterns. An alternative idea is not to set a specific length but to instead increase the length, allowing for mismatches as one proceeds, until one gets to a combination of length and number of mismatches that is statistically significant. This can be performed using a suffix tree (or suffix array), an efficient means of storing all of the subsequences, of all lengths, that occur in a sequence set. Because it only keeps track of the words that occur in the sequences, the total number of patterns to be explored is limited by the total sequence length rather than the number of possible words of a given length. For example, in Figure 7-5A are 22 sequences, each ~200 bp long (counting both strands), so the total number of possible sites is only ~4400, less than the number of 7-long sequences (so not every 7-long sequence can occur in the set). Programs such as WEEDER work in a similar manner, tracking the occurrences of words and allowing mismatches, to find the most significant consensus sequences (Pavesi et al. 2001).

Greedy Alignments

A very simple algorithm to find the best PWM is to create alignments of sites from the sequences, beginning with pairwise alignments and adding new sites until every sequence has contributed a site. This is a "greedy algorithm," in that at each step it saves the best partial solutions from the previous step and eliminates the remainder (Stormo and Hartzell 1989; Hertz and Stormo 1999). A greedy algorithm could throw out the one that will lead to the overall optimum, so there is no guarantee that it will find the best. In fact, if only the single best alignment is saved at each step, it is highly probable that the algorithm will not find the optimal alignment. Thus, generally the algorithm works by saving, at each step, some number (anywhere from thousands to millions) of the best alignments and continually adding new sequences until a final answer is obtained. The objective function is the IC, although in more recent versions this is replaced by an E-value calculation, which is the number of alignments with equal or higher information content expected by chance. The lower the E value, the more significant the alignment. The algorithm can be easily modified to allow some sequences to be devoid of a binding site, anticipating that there may be some "bad data" in the set, as well as to allow some sequences to contribute more than one binding site. In each case, the calculation of the E value is adjusted to account for the number of possible alignments with the given constraints. Other constraints can also be added, such as searching only for symmetric patterns if it is known that the TF binds as a homodimer. Any other prior information about the pattern's possible look can be used to influence the choice of "best alignments" at each step of the process. In general, this approach can be a very effective means of finding the most significant PWM from a set of sequences such as those in Figure 7-5A.

Expectation Maximization

The expectation maximization (EM) approach also works to find a PWM by modeling the binding sites as coming from multinomial distribution (a probability of each base at each position in the binding sites; essentially, the PFM described above) and, simultaneously, a background distribution ($F(b,0)$) for the sequences not included in the sites (Lawrence and Reilly 1990). This approach treats the location of the site in each sequence as missing data, and it does an iterative updating of all of the model parameters to maximize the probability of the entire set of sequences. The strategy can be described by the following steps:

1. Given a PFM (Fig. 7-2C), calculate the probability of each position in a sequence being the binding site. Normalize the probabilities so that they sum to 1 across the sequence. The initial PFM can be obtained by assuming every position in each sequence is equally likely to be the binding site.

2. For each sequence, create the PFM obtained by summing together each site within the sequence weighted by its probability. The total PFM is obtained by averaging the PFMs from each sequence. The $F(b,0)$ values are obtained from every position in every sequence weighted by the probability that it is not included in the binding site.

3. Iterate those two steps until convergence: The PFM and the probabilities of the binding sites no longer change (or the amount of change is below some set value).

EM is guaranteed to converge; in fact, the total probability of the data is guaranteed to increase at each step until it reaches a maximum. It is not guaranteed to find the correct answer because it is completely deterministic and the final solution depends on the initial PFM. It is possible that the initial PFM leads to a local maximum of probability but not the global maximum. As with the greedy algorithm, the use of prior knowledge in determining the initial PFM may increase the likelihood of finding the global maximum. For example, instead of assuming initially that every position in each sequence is equally likely to be the binding site, one may know that binding sites are more likely to be near the right (or left or middle) of the sequences based on known positional biases for the type of pattern for which one is searching. Or, one may have a general idea of what the pattern may look like, based on the type of TF involved, and the initial PFM could be biased toward that pattern. If one suspects that there may be sequences without binding sites, or sequences with multiple binding sites, those expectations can be included in the overall probability calculations. EM can also be used to discover combinations of patterns simultaneously if, for example, it is expected that there may be two TFs that bind coordinately to the set of sequences.

Gibbs Sampling

Gibbs sampling is similar in some ways to EM but different in important respects. One important difference is that the total probability is guaranteed to increase at each step with EM but with Gibbs sampling, it may decrease (Lawrence et al. 1993). This allows the algorithm to escape the local maxima in which EM may get trapped and possibly find a better solution. It also means that different runs from the same starting conditions may result in different answers; therefore, it is prudent to make several runs and compare the results. If the results are all the same, or nearly so, this gives added confidence that these results are the best possible solution. The PFM in Gibbs sampling is obtained from a single site in each sequence instead of the probability-weighted average of all possible sites used in EM. The site chosen from each sequence is obtained by a sampling strategy in which every site is weighted by the ratio of its PFM probability to the background probability. The strategy can be described by the following steps:

1. Given N input sequences, choose a site from $N - 1$ of these and make a PFM from those sites (because the sample size is only $N - 1$ sites, pseudocounts are used to prevent any elements from being assigned 0 probability; see Fig. 7-2D). In addition, determine the background probability $F(b,0)$ of the data set (or this can be defined independently). The initial PFM may be chosen randomly from $N - 1$ sequences, although prior knowledge could also be used in the selection of the initial set of putative sites.

2. Given the PFM, select one site from the sequence that was not included in the previous step. The probability of each site being selected is proportional to the ratio of its PFM probability to its background probability.

3. Pick one sequence from the previous set and remove its site from the previous PFM. Add the new site and update the PFM.

4. Iterate steps 2 and 3 either for a fixed number of cycles or until the total probability of the data appears to plateau.

If the resulting PFM is nearly identical after several runs of the program, that PFM is likely to be the true optimum solution. If each run gives a different PFM, the one with the highest total probability could be selected as the best result, but there is the risk that additional runs could find even better solutions. It is also worth determining an E value for each PFM because it is possible that none of the results are significant. Running the Gibbs sampler program on shuffled sequences that maintain the same background probabilities but no longer contain any conserved motif can be used to estimate the distribution of PFMs that can be expected by chance.

PHYLOGENETIC FOOTPRINTING

Another approach to identifying regulatory sites is to assume that a given gene is likely to be regulated by the same factors in closely related species. One can take the presumed regulatory regions, for instance the upstream regions where the promoter and at least some of the regulatory sites are likely to be located, of several species and identify subregions that are highly conserved by using a multiple sequence alignment method. For example, in Figure 7-6B, several sequences from a reference genome of interest, represented with bold lines, are compared to orthologous regions from three other species, and highly conserved regions are found, indicated with gray boxes. This procedure is called "phylogenetic footprinting" and has been combined with the same methods described above for motif discovery (Wasserman et al. 2000; Wang and Stormo 2003; Sinha et al. 2004; Siddharthan et al. 2005). Those regions are mostly likely to contain the regulatory sites because they tend to be less tolerant of mutation and therefore evolve more slowly than nonfunctional regions of the genome. By combining motif discovery with phylogenetic footprinting, motifs that are both conserved through evolution and occur in the coregulated genes are found. This can provide added confidence that the motif is the regulatory site of interest and several such sites can be identified through iterative use of the same steps.

This chapter has presented methods for modeling and predicting TF-binding sites, both when collections of sites are known and when they must be discovered from incomplete data. These approaches complement structural information about binding and thermodynamic analyses of binding affinities. The information from the different approaches must be consistent and each perspective provides insights not available from the others. Chapter 8 also describes bioinformatics approaches but focuses on the protein side of the interaction. It describes how TFs can be identified from

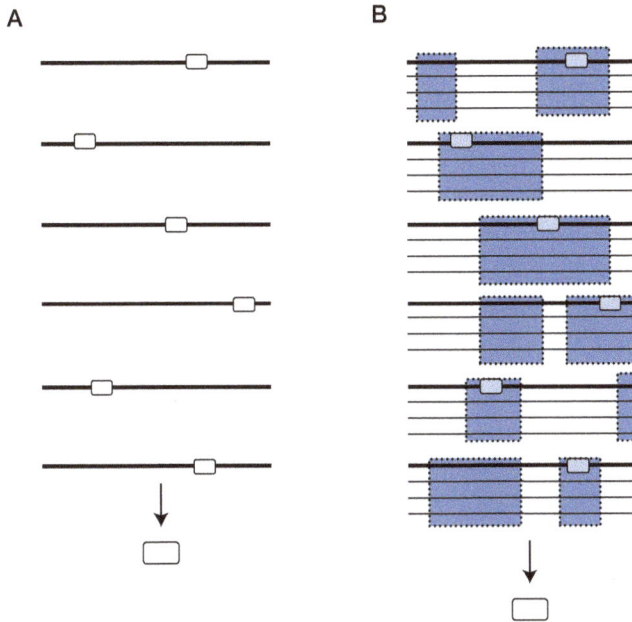

Figure 7-6. Use of phylogenetic conservation in motif discovery. (*A*) Set of sequences, from one species, to be used for motif discovery. The lines represent genomic segments, as in Figure 7-5A, and the boxes are the binding sites (initially unknown and to be discovered). This part of the figure corresponds to the motif discovery problem of Figure 7-5. (*B*) Each sequence from *A* is aligned with the orthologous segments from three other species (thin lines aligned with bold lines from *A*). Conserved segments are included in the shaded boxes. By confining the motif discovery search to the conserved regions, it should be easier to find the correct sites, and if the sites themselves are conserved, this provides added confidence to the predicted motif.

genomic sequences and how their similarity to other factors informs how they interact with DNA. It also describes bioinformatics approaches to predicting the binding specificity of TFs based only on their protein sequences and how that information can be used to design proteins with desired specificity.

REFERENCES

Djordjevic M, Sengupta AM, Shraiman BI. 2003. A biophysical approach to transcription factor binding site discovery. *Genome Res* **13:** 2381–2390.

Galas DJ, Eggert M, Waterman MS. 1985. Rigorous pattern-recognition methods for DNA sequences: Analysis of promoter sequences from *Escherichia coli*. *J Mol Biol* **186:** 117–128.

Hertz GZ, Stormo GD. 1999. Identifying DNA and protein patterns with statistically significant alignments of multiple sequences. *Bioinformatics* **15:** 563–577.

Lawrence CE, Reilly AA. 1990. An expectation maximization (EM) algorithm for the identification and characterization of common sites in unaligned biopolymer sequences. *Proteins* **7:** 41–51.

Lawrence CE, Altschul SF, Boguski MS, Liu JS, Neuwald AF, Wootton JC. 1993. Detecting subtle sequence signals: A Gibbs sampling strategy for multiple alignment. *Science* **262:** 208–214.

Pavesi G, Mauri G, Pesole G. 2001. An algorithm for finding signals of unknown length in DNA sequences. *Bioinformatics* **17:** S207–S214.

Schneider TD, Stephens RM. 1990. Sequence logos: A new way to display consensus sequence. *Nucleic Acids Res* **18:** 6097–6100.

Schneider TD, Stormo GD, Gold L, Ehrenfeucht A. 1986. Information content of binding sites on nucleotide sequences. *J Mol Biol* **188:** 415–431.

Siddharthan R, Siggia ED, van Nimwegen E. 2005. PhyloGibbs: A Gibbs sampling motif finder that incorporates phylogeny. *PLoS Comput Biol* **1:** 534–556.

Sinha S, Blanchette M, Tompa M. 2004. PhyME: A probabilistic algorithm for finding motifs in sets of orthologous sequences. *BMC Bioinformatics* **5:** 170.

Stormo GD, Hartzell GW 3rd. 1989. Identifying protein-binding sites from unaligned DNA fragments. *Proc Natl Acad Sci* **86:** 1183–1187.

Stormo GD, Schneider TD, Gold L, Ehrenfeucht A. 1982. Use of the "perceptron" algorithm to distinguish translational initiation sites in *E. coli. Nucleic Acids Res* **10:** 2997–3011.

Wang T, Stormo GD. 2003. Combining phylogenetic data with coregulated genes to identify regulatory motifs. *Bioinformatics* **19:** 2369–2380.

Wasserman WW, Palumbo M, Thompson W, Fickett JW, Lawrence CE. 2000. Human-mouse genome comparisons to locate regulatory sites. *Nat Genet* **26:** 225–228.

Zhao Y, Granas D, Stormo GD. 2009. Inferring binding energies from selected binding sites. *PLoS Comput Biol* **5:** e1000590.

Zhao Y, Ruan S, Pandey M, Stormo GD. 2012. Improved models for transcription factor binding site identification using nonindependent interactions. *Genetics* **191:** 781–790.

FURTHER READING

The following offer reviews on DNA motifs and motif-finding methods:

Stormo GD. 2000. DNA binding sites: Representation and discovery. *Bioinformatics* **16:** 16–23.

Lässig M. 2007. From biophysics to evolutionary genetics: Statistical aspects of gene regulation. *BMC Bioinformatics* **8:** S7.

van Nimwegen E. 2007. Finding regulatory elements and regulatory motifs: A general probabilistic framework. *BMC Bioinformatics* **8:** S4.

CHAPTER 8

Bioinformatics of Transcription Factors and Recognition Models

CHAPTER 3 DESCRIBES THE STRUCTURES OF MANY DIFFERENT families of transcription factors and the various modes of interaction with DNA that are used. Chapter 4 provides a look into the protein–DNA interface for many proteins to see how they can distinguish among different sequences. This chapter describes how transcription factors (TFs) can be identified on the basis of their protein sequences and classified into different protein families. Also described is how information about the specificity of specific families, using the experimental methods described in Chapter 5 and the bioinformatics approaches described in Chapter 7, can lead to recognition models for protein–DNA specificity. Those recognition models can be used to predict the specificity of a TF based on its protein sequence and aid in the design of TFs with desired specificities.

IDENTIFYING HOMOLOGOUS TFs

When new DNA sequences are obtained, one of the first objectives is to identify the proteins that are encoded in the DNA and predict their functions. The best methods for predicting protein functions are based on evolutionary processes during which amino acid substitutions accumulate over time, leaving unchanged those that are required for function. Two methods are most often used to predict the function of a protein. In the first method, the new protein sequence is compared to a database of all known proteins to find those that are significantly similar according to the evolutionary models of substitution probabilities. An alternative approach, described in the next section, uses the constraints on amino acid distributions at each position that are characteristic of particular protein families.

As shown in Chapter 3, TFs can be classified into distinct protein families on the basis of their overall structure and how they bind DNA. Structures are known for only a very small fraction of all TFs, but they can be classified into families based on their sequence similarities. An assortment of methods are available for measuring the similarity between two protein sequences, but the most reliable methods are those based on models of evolutionary processes. The DNA that encodes a protein can undergo a variety of mutations, but the most common are base substitutions (replacing one base pair of the gene with a different base pair) and small insertions and deletions (indels). Most changes that are not deleterious are selectively neutral, but occasionally beneficial mutations can improve the function of the protein. In either case, mutations accumulate over time so that diverse protein sequences can perform the same function in different species, and the more distantly related the species, the more divergent their gene sequences. In addition, within a species, a gene can be duplicated and then accumulate mutations to take on related but somewhat different functions in the cell. Genes that are related by a common ancestor are homologous, often referred to as homologs, and two classes can be distinguished. Orthologs are genes in different species that derive from a common ancestor and provide the same, or very similar, function in each species. Paralogs are genes that are duplicated within a species and take on slightly different roles. Paralogs are a very common means for a genome to attain a diversity of TFs that can generate a wide variety of expression patterns. In humans, the two most common TF families are zinc fingers, with more than 800 members, and homeodomains, with more than 200 members. Although it is possible that some of these members could be redundant, most are likely to have alternative functions, such as binding to alternative sites, interacting with alternative factors, and responding to alternative signals.

Both paralogs and orthologs are homologous proteins and if their sequences have not diverged too much they can generally be recognized by sequence comparisons that account for evolutionary processes. This requires a model for the probabilities of different amino acid substitutions, and a commonly used model is the BLOSUM62 substitution matrix (Fig. 8-1A) (Henikoff and Henikoff 1992). This matrix is based on a large number of aligned, homologous proteins, and the score for each amino acid pair is

$$S(a_i, a_j) = \frac{1}{\lambda} \log \frac{\Pr(a_i a_j)}{\Pr(a_i)\Pr(a_j)}, \tag{8-1}$$

where a_i and a_j are the two amino acids (which may be the same). The probability in the numerator is for both amino acids occurring in the same aligned position, and the probabilities in the denominator are for their independent occurrences. λ is a scaling factor, dependent on the base of the logarithm. The "62" in BLOSUM62 refers to the fact that the values come from aligned sequences that are <62% identical, which has been found to be a good choice for general database searches. When searching for very highly diverged sequences, it may be better to use a scoring matrix built from less similar sequences, such as BLOSUM45. Each is a log-odds matrix, similar to those described in Chapter 7, and if the two amino acids occur independently of one another,

A   Log-odds matrix for BLOSUM62

	A	C	D	E	F	G	H	I	K	L	M	N	P	Q	R	S	T	V	W	Y
A	4	0	-2	-1	-2	0	-2	-1	-1	-1	-1	-2	-1	-1	-1	1	0	0	-3	-2
C		9	-3	-4	-2	-3	-3	-1	-3	-1	-1	-3	-3	-3	-3	-1	-1	-1	-2	-2
D			6	2	-3	-1	-1	-3	-1	-4	-3	1	-1	0	-2	0	-1	-3	-4	-3
E				5	-3	-2	0	-3	1	-3	-2	0	-1	2	0	0	-1	-2	-3	-2
F					6	-3	-1	0	-3	0	0	-3	-4	-3	-3	-2	-2	-1	1	3
G						6	-2	-4	-2	-4	-3	0	-2	-2	-2	0	-2	-3	-2	-3
H							8	-3	-1	-3	-2	1	-2	0	0	-1	-2	-3	-2	2
I								4	-3	2	1	-3	-3	-3	-3	-2	-1	3	-3	-1
K									5	-2	-1	0	-1	1	2	0	-1	-2	-3	-2
L										4	2	-3	-3	-2	-2	-2	-1	1	-2	-1
M											5	-2	-2	0	-1	-1	-1	1	-1	-1
N												6	-2	0	0	1	0	-3	-4	-2
P													7	-1	-2	-1	-1	-2	-4	-3
Q														5	1	0	-1	-2	-2	-1
R															5	-1	-1	-3	-3	-2
S																4	1	-2	-3	-2
T																	5	0	-2	-2
V																		4	-3	-1
W																			11	2
Y																				7

B

```
...Y K  C  E  F  A  D  C  E  K  A  F  S  N  A  S  D  R  A  K  H  Q  N  R  T  H...
...Y Q  C  P  -  -  D  C  Q  K  S  Y  S  T  F  S  G  L  T  K  H  Q  Q  F  -  H...
+7+1 +9 -1 -7 -1 +6 +9 +2 +5 +1 +3 +4 +0 -2 +4 -1 -2 +0 +5 +8 +5 +0 -3 -7 +8
=59
```

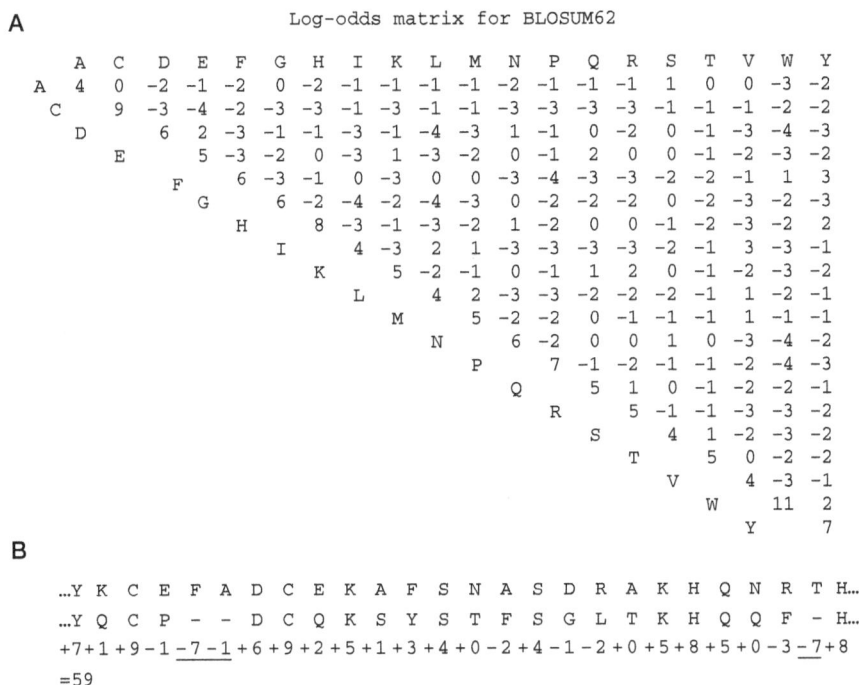

Figure 8-1. Measuring protein similarities. (A) BLOSUM62 matrix of amino acid similarities. Each element is the score between the alignment of two amino acids (the diagonal is the score for an amino acid with itself). (B) Score for the alignment of two (partial) protein sequences. Below each pair of aligned amino acids is the score from the matrix in A. Indels, " − ," are given a score of −7 for the occurrence of a gap and −1 for each additional position in the gap beyond the first (underlined).

their observed co-occurrence probability will approximately equal their expected co-occurrence probability, which is just the product of their independent probabilities. A score of 0 indicates that the pair of amino acids occurs in the same position with the frequency expected by chance, and positive (negative) scores are for pairs that occur more (less) often than expected randomly.

Examination of the BLOSUM62 matrix (Fig. 8-1A) shows that scores for every amino acid with itself (on the diagonal) are positive and are always the largest value in every row and column. This is because having no substitution is the most likely occurrence in related proteins, which is true even at large divergences such as in the BLOSUM45 matrix. Although most matrix values are negative, a few are positive, indicating that those substitutions occur more frequently than expected by chance, given the overall amino acid composition. For example, the set of amino acids I, L, M, and V (Ile, Leu, Met, and Val) all have positive scores with one another. They are all large hydrophobic amino acids (see Fig. 3-2B), and substitution of one for another will often not have a negative effect on the protein structure or function.

On the other hand, P (Pro) is negative with every other amino acid substitution, consistent with its unique properties (see Fig. 3-2B) and the common deleterious effects on structure or function if it replaces another amino acid. Scores between any two amino acids are based on empirical evidence of their interchangeability but serve as a general measure of similarity in their properties.

To assess whether two proteins are homologous, the score of their alignment is calculated, based on the scoring matrix, and then it is determined whether that score is significantly higher than expected for nonhomologous proteins. Figure 8-1B shows an alignment of two zinc finger domains and the score for each aligned pair of amino acids based on the BLOSUM62 matrix. There are also three indel positions or "gaps," indicated by " − " symbols, where one protein has an amino acid that is missing from the other protein. Two types of gap scores are used: one for the occurrence of a gap and a separate one based on its length. This approach is based on the empirical observation that although indels are relatively rare, it is more likely to have two (or more) adjacent to one another than to have the same number occurring independently, presumably because a single mutational event can remove (or insert) several amino acids at once, and that is more likely than having multiple independent events that each remove (or insert) a single amino acid. It has been determined that for the BLOSUM62 scoring matrix the best gap opening score is −7 and the gap extension score is −1. Thus, the second gap in the alignment, which is only one position long, has a total score of −7, whereas the first gap, which is two positions long, has a total score of −8. The complete alignment is 26 positions long (the length of the longer sequence that contains no gaps in the alignment) and the two sequences are 38% identical (10/26 positions) and 58% similar (15/26 positions have scores >0). The inference about homology is based on the total score of the alignment, which is 59. Of course, there are many possible alignments between two proteins when indels are allowed, so first, one must be assured that the alignment being evaluated has the highest possible score. Dynamic programming methods, such as the Smith–Waterman algorithm (1981), can find the highest-scoring alignment in an amount of time that is proportional to the product of the length of the two sequences (see Box 8-1). That time is adequate for comparing a small number of proteins, but to search the entire database of known protein sequences with every newly sequenced protein would be prohibitively time consuming. Therefore, fast database search methods, such as BLAST (Altschul et al. 1990) have been developed that greatly reduce the time required to locate all of the highest-scoring proteins in the database.

BLAST uses an approximation to the full dynamic programming algorithm, and consequently, it cannot be guaranteed to find every protein in the database with a score above a given threshold. However, current versions are very sensitive, and the probability of missing a high-scoring homolog is quite low. BLAST is by far the most commonly used method because of its speed, sensitivity, and convenience of use. Furthermore, rigorous statistical methods have been developed to calculate the significance of any score associated with an alignment of protein pairs. The significance is often

Box 8-1. Optimal Alignments with Dynamic Programming

The best possible alignment between two sequences can be obtained by a procedure known as dynamic programming, which is described in the figures below. First, one needs a scoring matrix that provides a score to the alignment of any two amino acids. For this example, we use the BLOSUM62 matrix provided in Figure 8-1A. One also needs a penalty for deletions/insertions (indels), which are amino acids that occur in only one of the sequences (and are indicated by " − " or "gap" in the alignment) as shown in Figure 8-1B. Here, we use −7 as the indel penalty, but we use only a single penalty rather than separate gap opening and gap extension penalties. Having a separate gap extension penalty increases the complexity of the algorithm slightly and is thus omitted for this brief description.

We want to align two seven-long protein sequences: RYLNAKY with RFHARFL. We start by placing the two sequences on the edges of the matrix; the elements within the matrix will be the best scoring alignment up to that point, as described in the algorithm below.

	−	R	Y	L	N	A	K	Y
−								
R								
F								
H								
A								
R								
F								
L								

The matrix also contains an initial row and column for leading gaps in the alignment, where one protein would extend beyond the other at the beginning of the alignment. The matrix elements for the initial row and column are easily calculated because it is just the indel penalty (−7) for each amino acid in one sequence that occurs before the first amino of the other sequence:

	−	R	Y	L	N	A	K	Y
−	0	−7	−14	−21	−28	−35	−42	−49
R	−7							
F	−14							
H	−21							
A	−28							
R	−35							
F	−42							
L	−49							

Box 8-1. (*Continued.*)

The rest of the matrix is filled in, following these simple rules:

$$M(i,j) = \max \left\{ \begin{array}{c} M(i-1,j-1) + S(a_i,b_j) \\ M(i-1,j) + \delta \\ M(i,j-1) + \delta \end{array} \right\} \quad \delta \text{ is the "indel" penalty.}$$

The value of any element in the matrix, say, $M(i,j)$, is determined by the largest of the three possible values for that element, which is determined by the three elements above and/or to the left of it as well as the similarity score for the amino acids in the two proteins that occur at positions i and j.

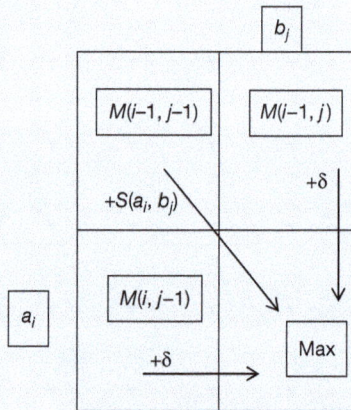

Filling in all of the elements of the matrix following that procedure leads to the following alignment matrix (left as an exercise for the reader):

	–	R	Y	L	N	A	K	Y
–	0	–7	–14	–21	–28	–35	–42	–49
R	–7	5	–2	–9	–16	–23	–30	–37
F	–14	–2	8	1	–6	–13	–20	–27
H	–21	–9	1	5	2	–5	–12	–18
A	–28	–16	–6	0	3	6	–1	–8
R	–35	–23	–13	–7	0	2	8	1
F	–42	–30	–20	–13	–7	–2	1	11
L	–49	–37	–27	–16	–14	–8	–4	4

The highest alignment score for those two sequences can be found in the lower right corner of the matrix, which is 4 in this example. The actual alignment that produces that score can be found by tracing back from the lower right corner and identifying the preceding

Box 8-1. (*Continued.*)

element that gave rise to the score, shown in large bold numbers in the matrix. This trace-back path determines the optimum alignment:

```
RYLNAKY -
RF - HARFL
5 + 3 - 7 + 1 + 4 + 2 + 3 - 7 = 4.
```

That procedure has determined the optimal "global alignment": the best scoring end-to-end alignment for the two sequences. Often, it is more useful to determine the optimum "local alignment," which is the highest-scoring segments within the two sequences. This can be done using the Smith–Waterman algorithm, which requires only one modification to the rules for filling in the alignment matrix: No negative score is allowed. Anytime an element would be negative, it is replaced by a 0:

$$M(i, j) = \max \left\{ \begin{array}{c} M(i-1, j-1) + S(a_i, b_j) \\ M(i-1, j) + \delta \\ M(i, j-1) + \delta \\ 0 \end{array} \right\}.$$

This eliminates the need for the initial row and column containing gaps and it also means that the best local alignment is obtained by the highest score anywhere in the matrix—not necessarily the element in the lower right corner. Trace back from that highest element until the last nonzero is reached, which determines the start of the alignment. The matrix for the best local alignment of the two sequences used above is now

	R	Y	L	N	A	K	Y
R	5	0	0	0	0	0	0
F	0	8	1	0	0	0	3
H	0	1	5	2	0	0	0
A	0	0	0	3	6	0	0
R	5	0	0	0	2	8	1
F	0	8	1	0	0	1	11
L	0	0	12	5	0	0	4

And the best local alignment, with a score of 12, is

```
RYL
RFL
5 + 3 + 4 = 12
```

This is slightly better than the score of 11 obtained by leaving off the last position of the best global alignment shown above.

reported as an E value, which is the expected number of matches in the entire database with a score as high as or higher than that observed. The E value is calculated by

$$E = Kmne^{-\lambda S}, \tag{8-2}$$

where S is the score of the alignment, λ is the scaling factor described above, m and n are lengths of the query sequence and the total database, respectively, and K is a constant that depends on the scoring matrix and the overall composition of the database. The lower the E value, the more significant the alignment between the two proteins. Note that because most elements of the scoring matrix are negative, a random alignment of any two proteins is expected to have a negative total score. But because of the total number of possible alignments, related to the product of m and n, a rather large score is required before two proteins can reliably be inferred as homologous.

HIDDEN MARKOV MODELS OF PROTEIN FAMILIES

One limitation of pairwise comparison methods, such as BLAST, is that all positions within the protein are scored using the same matrix, as if they are all equally mutable. But some positions within a protein may be critical for its proper structure or function and may tolerate only a limited set of substitutions. In some cases, only a single amino acid is allowed at a critical position in a protein for it to be functional. In such cases, homologous proteins will have the same constraints on their sequences, and they can be more reliably identified by taking those family-specific constraints into account. Instead of using a general scoring matrix (such as BLOSUM matrices), which represents the average substitution probabilities over all protein positions, more sensitive detection of homologous proteins can be obtained by using position-specific scoring methods that model common characteristics of each particular protein family.

A specific form of hidden Markov model (HMM), called a profile HMM (Krogh et al. 1994; Eddy 1998), provides a convenient and useful mathematical representation that captures well the similarity between members of a protein family. Databases such as Pfam determine profile HMMs for many protein families and serve as very useful tools for recognizing new members of a specific family as well as describing important characteristics of the family. It is easiest to understand a profile HMM using an example. Figure 8-2A shows an alignment of protein sequences from the C2H2

Figure 8-2. Modeling protein sequence alignments. (A) Alignment of 20 zinc finger proteins, taken from a total of 195 proteins in the Pfam database (Pfam family zf-C2H2; Pfam ID: PF00096). (B) General structure of a profile hidden Markov model (profile HMM). (Triangles) Begin and end states for the HMM, (rectangles) match states that correspond to positions in the protein alignments with a minority of gaps, (circles) delete states that represent amino acids missing in particular proteins, (diamonds) insert states where some proteins have extra amino acids occurring between match states. (*Legend continues on facing page.*)

A

```
YKCEF-ADCE---KAFSNA---SDRAKHQNR--TH
YQCP---DCQ---KSYSTF---SGLTKHQQF---H
HRCPH-SNCT---RSFAMR---ESLVRHLVV---H
LRCS---RCS---KQFNHP---TELVQHEKV--LC
HSCP---KCG---KRFKRR---DHVLQHLNKKIPC
TDCRW-DGCS---QEFDSQ---EQLVHHINS--EH
DDCQ---DCY---ETLTSS---FEVIVHRINH-HH
LYCKW-DNCG---MIFNQP---ELLYNHLCH--DH
FQCD---ICK---KTFKNA---CSVKIHHKN--MH
VQCS---ICF---KTFCDK---GALKIHFSA--VH
LQCP---KCP---FVTEFK---HHLEYHIRK---H
YPCRKDSTCP---FVGKTW---SDYMKHAAE--LH
WKCGK-KDCG---KMFARK---RQIQKHMKR---H
YSCA---DCG---KHFSEK---MYLQFHQKNPSEC
IPCH---ICG---EMFSSQ---EVLERHIKAD-TC
AQCP---ICQ---QFYPLK---ALEKTHLDE---C
YTCG---YCTEDSPSFPRP---SLLESHISL--MH
HYCP---MCH---QQFERP---QHVADHMQL--CH
KQCR---YCP---KSFSRP---VNTLRHMRM---H
YQCQ---LCK---KSISRHGQYANLLNHLSR---H
```

C

B

D

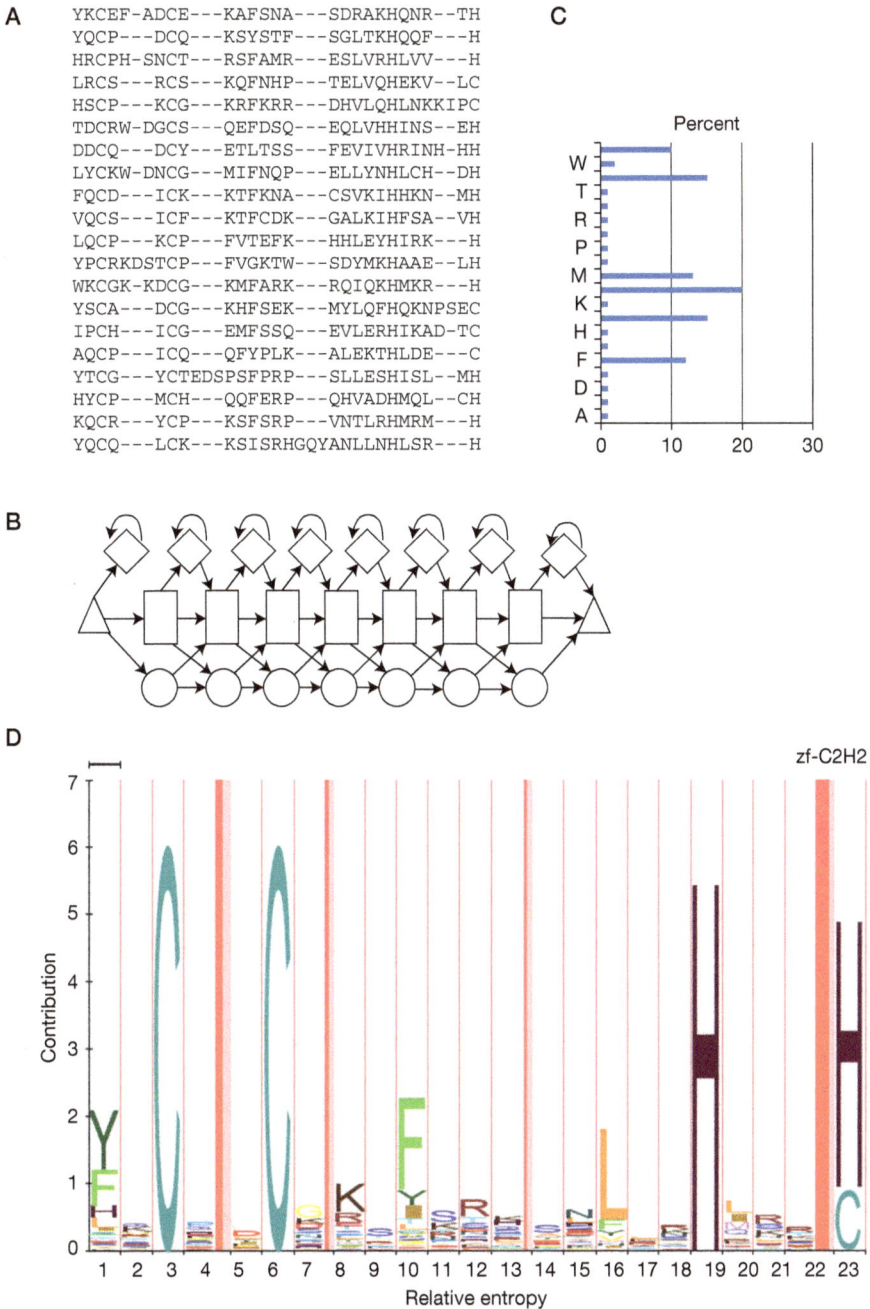

Figure 8-2. (*Continued.*) (*C*) Graphical representation of the match state probabilities for a particular position, displayed as percentages. Pseudocounts are added to each amino acid so that no probabilities are 0. (*D*) Sequence logo for the zinc finger family of *A*, but based on the entire 195 sequences. Logos for the match states are made as described in Chapter 7. Insertion states are represented by vertical bars with their widths proportional to their frequency. The left, darker portion of the bars corresponds to the frequency of inserts occurring at that position, and the right, lighter portion corresponds to their length.

zinc finger family, just showing a partial sequence from each protein: the zinc finger domain containing the critical 2 Cys (C) and 2 His (H) residues that coordinate the zinc ion and give the family its name (see Fig. 3-6A). Three of those four positions are absolutely conserved in this alignment: the two Cys residues and the first of the aligned His residues; the last His residue is sometimes substituted with another Cys. None of the other positions in the alignment are completely conserved, and the degree of variability differs in different positions. For example, a position (the first R in the first sequence) is nearly always a Leu (L in one-letter code) but can be substituted most commonly by other large hydrophobic amino acids such as L, I, and Y. Some of the aligned positions contain gaps, where an amino acid is missing in that particular protein but occurs in other members of the family. A column that contains mostly amino acids and fewer (or no) gaps is referred to as a "match position" (or a "match state" in the profile HMM) and can be used for the position-specific amino acid probability distribution that helps to define the family. A gap within a match position indicates a protein that has undergone a deletion relative to most proteins in the family. The alignment of Figure 8-2A contains a total of 35 positions, 24 of which are match positions and 11 of which are insert positions. One match position—the next to last position in the alignment—contains deletions that occur in seven of the 20 sequences in this sample.

Figure 8-2B shows the general features of a profile HMM that can be used to represent any protein family. There are three types of "states" indicated by the different shapes: rectangles, circles, and diamonds (the triangles are special "begin" and "end" states for the model). The rectangles are the "match states," or positions in the alignment that occur in most proteins. Each match state has a probability distribution of amino acids associated with it, often represented graphically as a set of bars—one for each amino acid—with their lengths proportional to the probability for that amino acid occurring at that position (Fig. 8-2C). The diamonds represent insert states, or positions that occur in a minority of the protein sequences in the family. They could also have a probability distribution over the amino acids that occur in those states, but because they are often quite rare (maybe only one or very few members of the family have a particular insertion), it may be difficult to estimate that distribution reliably; thus, usually all insert states are assigned a common distribution of either all amino acids equally likely or with probabilities that are the average over all protein sequences. The circles represent delete states, where an amino acid that occurs in most members of the family is missing in a particular member of the family.

The arrows in the profile HMM represent the allowed transitions between states, and each transition has a probability associated with it, often represented graphically by the thickness of the line. The delete states are aligned with the match states because they refer to deleting a specific match state. A deletion of multiple amino acids in a particular protein corresponds to transitions from one delete state to the next, until all of the deleted positions have been included. The insert states occur between match states, and only insert states allow transitions back to the same state, which corresponds to multiple amino acids being inserted in a particular protein (such as several examples in Fig. 8-2A). Except for the return transition allowed for insert states, all other transitions are forward, moving to the next match, insert, or delete state. Transitions are not allowed between insert and delete states; those can only transition to another state of the same type or to a match state. This corresponds to the observation that substitutions are more common than indels, so rather than allow an amino acid in a protein sequence to be considered an insertion following (or preceding) immediately from a deletion, it should be considered to be a match position, even if that amino acid has a low probability of occurrence at that match state. Importantly, all of the transition probabilities from any state must sum to 1, just as the amino acid distributions within a match or insert state must sum to 1.

The match state amino acid probabilities could be assigned their observed frequencies in the aligned sequences, but this runs the risk of assigning 0 probability to an amino that might exist in that position but has not been observed in the current sample. This risk is quite large if the sample is small. For example, the sample shown in Figure 8-2A contains only 20 sequences, so it would be exceedingly unlikely to observe all 20 amino acids; each would have to occur exactly once. Therefore, assigning a probability of 0 to an unobserved amino acid in that case has a high risk of being incorrect and overfitting to a small sample. On the other hand, if the sample size were quite large, say 2000 proteins, an unobserved amino acid may well have 0 probability of occurring at that position. Adding a fixed total of pseudocounts to the observed amino acid counts at each position deals appropriately with both situations, as in the use of pseudocounts for position weight matrices (see Fig. 7-2). If the sample is small, the pseudocounts can have a large influence on assigned probabilities; if the sample is large, the pseudocounts affect only a small change over the observed frequencies but still disallow a probability of exactly 0. A simple method is to add a pseudocount of 1 for each amino acid at each position, but other choices are possible and may be more appropriate. One can adjust the pseudocounts for each amino acid based on the observed amino acids to account for the fact that substitutions of similar amino types are more likely than dissimilar ones. For example, aligned position 25 of Figure 8-2A (the first R of the first sequence, described above) is mostly L, with a common substitution of V, another large hydrophobic amino acid. V and I are also large hydrophobics and are observed along with a few nonhydrophobic amino acids. But F and M are not observed in this sample, even though they are large and hydrophobic and one might expect them to substitute well. In fact, in the alignment of 195 proteins of the family from which this sample of 20 was drawn, the five most common amino acids are L>F>Y>V>M. Thus, pseudocounts biased toward the properties of the observed amino acids in the small sample provide a better overall model for the entire family.

Figure 8-2D is another graphical representation of a profile HMM for the C2H2 zinc finger family. This is a sequence logo, similar to the one described in Chapter 7, where the height of the column at each position is the information content (or relative entropy) of that aligned position, and the height of each letter within a column is proportional to its probability at that position. One difference between these logos and those for binding sites that we saw before is the inclusion of insert states—the vertical red two-toned bars. The widths of the bars are proportional to total frequency of inserts, with the left (darker) portion the probability of at least one amino acid insert. The binding-site position frequency matrix (PFM) models of Chapter 7 can be considered to be profile HMMs without any insert of delete states, because protein-binding sites on DNA sequences are generally all the same length and do not require indels to align them. There are some cases where binding sites can be of variable length, requiring gaps in their alignment, and in such cases, profile HMMs can be used; but for most TFs, PFMs (or position weight matrices [PWMs]) are sufficient. There are a few discrepancies between the logo of Figure 8-2D and the alignment of Figure 8-2A. One is that there are only 23 match states in the logo, whereas we determined above that there were 24 match positions and 11 insert positions in the alignment. The logo is based on 195 members of the family. In that complete alignment, position 34, which had 13 matches and seven deletions in the small sample, has more deletions than amino acids and thus is considered an insert rather than a match state in the complete model. Other discrepancies occur where the probabilities in the total family are significantly different than that observed in the small sample, as in the example described above, where match state 16 (alignment position 25) has F as the second most common amino acid, even though it was not observed at all in the small sample.

The generative models described in Chapter 7 for binding sites are also a useful way to think about profile HMMs—as a machine that can generate sequences based on its internal probabilities: the "emission" probabilities for amino acids in the match and insert states and the transition probabilities between states. The probability of generating any particular sequence is the product of the probabilities of all of the steps—the transitions and emissions—that were used during its generation. Now, consider the reverse problem. Given a sequence, what is the probability of it being generated by the machine? A probability can be obtained by assigning each position in the sequence to a state in the profile HMM and then determining the product of the emission and transition probabilities that corresponds to those assignments. A dynamic programming algorithm, similar to the one described in Box 8-1 but used to align a sequence to a profile HMM instead of two sequences, can find the alignment that has the highest probability. It is also possible, using a different dynamic programming algorithm, to find the probability that the sequence was generated by the machine considering all of the different ways to align the sequence to the profile HMM. Often, there is a single alignment that is much better than all of the others, so just finding the best alignment is sufficient. But other times, there may be many different alignments with similar probabilities, for example, when there are different choices for delete or insert states with similar overall probabilities. In such cases, finding the total probability that the machine generated the sequence may provide a more meaningful assessment of whether the protein is a member of the family.

Knowing only the probability of a sequence being generated is not sufficient for deciding if the new protein is a member of the family represented by the profile HMM. All protein probabilities, even those for the most likely sequence, will generally be very low because they are the product of many numbers all less than 1, and many of them are usually much less than 1. A more useful approach is to compare the probability of the given sequence to the probability distribution over all sequences, or the expected probability of a random sequence. This is analogous to the odds ratios used in the PWMs of Chapter 7 and allows one to determine how much more (or less) likely the sequence is to have been generated by the profile HMM than is expected by chance. The probabilities of all known members of the family should also be determined. If the test sequence falls within or close to the range of known members, and if it is much higher than expected by chance, it is likely to be a true member of the family.

EXAMPLES OF TF PROFILE HMMs

Figure 8-2D shows the logo for the C2H2 zinc finger family whose structure bound to DNA was shown in Figure 3-6A. The two Cys (C) residues that coordinate the zinc ion are completely conserved at match positions 3 and 6. There are variable numbers of positions between them, as shown in Figure 8-2A, usually between 2 and 5. The first of the His (H) residues that coordinates the zinc is also completely conserved at match position 19, but the second His is sometimes replaced by Cys at match position 23. No other positions are close to being completely conserved, but match positions 1, 10, and 16 are primarily composed of large hydrophobic amino acids. By examining the structure, we learn that those positions are internal to the protein where they help to stabilize the structure. Figure 4-2A shows the interactions between amino acids for each of the three fingers of the protein zif268 with its consensus binding site. The interacting amino acids, numbered with respect to the α helix that fits in the major groove, are −1, 2, 3, and 6. Those amino acids correspond to match positions 12, 14, 15, and 18 (0 is not used in the structure numbering system, so there is only one match position, 13, between structure positions −1 and 2). Those positions are highly variable in the C2H2 zinc finger family, as expected because different protein sequences can be used to specify different binding sites. There is an overall preference for polar amino acids because they can form hydrogen bonds with DNA base pairs, but almost any protein sequence may be used.

Figure 8-3 displays logos for a number of different TF families and shows for each the variability at positions throughout the protein (in most cases, the logo and the profile HMM correspond to the DNA-binding domain; the complete proteins may be much larger and contain multiple different domains). Figure 8-3A is the GATA family, which includes C4 zinc finger proteins, and the four conserved Cys residues are readily seen. The W (Trp) at match position 12 is involved in structural stability of the protein, and the neighboring R (Arg) at position 13 is directly involved in binding to the

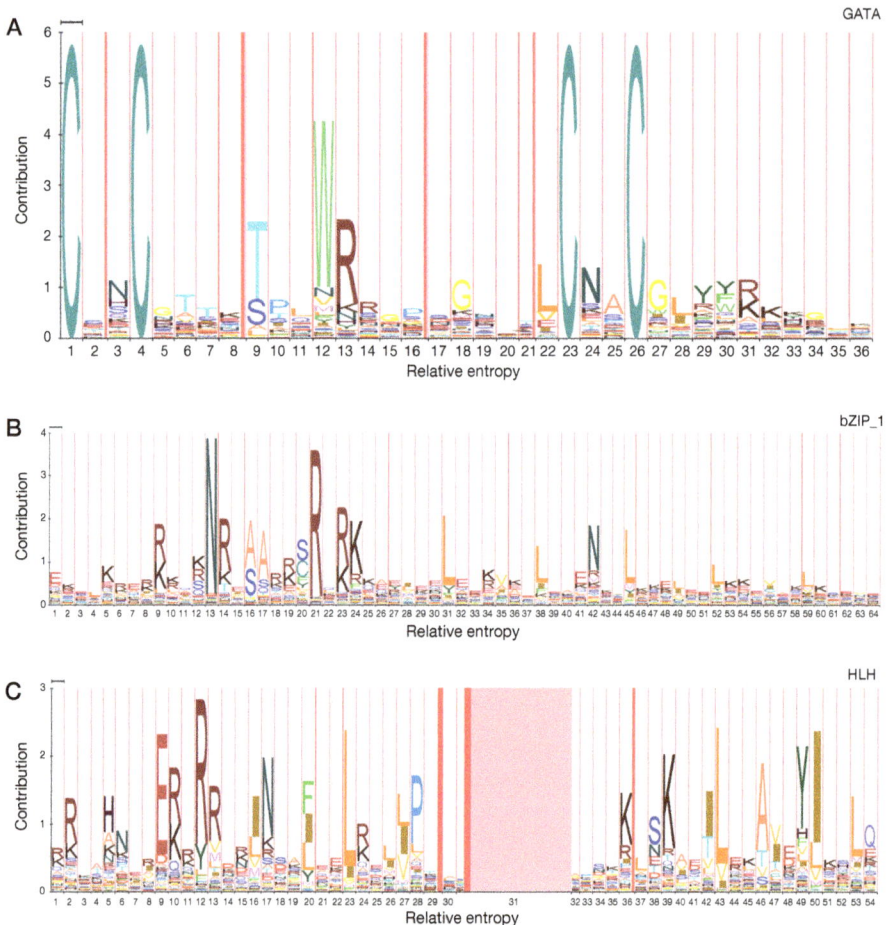

Figure 8-3. Pfam logos for common transcription factor families. (*A*) GATA transcription factors (GATA, PF00320; 74 sequences). (*B*) bZIP transcription factors (bZIP_1, PF00170; 22 sequences). (*C*) bHLH transcription factors (HLH, PF00010; 170 sequences). (*Legend continues on facing page.*)

conserved GATA binding sites, as are several of the other moderately conserved positions. Given that this family binds quite specifically to sites with GAT (the last A is less well conserved), it seems to be a bit surprising that the amino acids that provide specificity are not even more highly conserved. This highlights the fact that there are multiple ways to recognize a specific DNA sequence.

Figure 8-3B,C shows logos for the two classes of coiled helix proteins described in Chapter 3, the basic-region leucine-zipper (bZIP) family and the basic-region helix-loop-helix (bHLH) family. Both are characterized by an enrichment of basic amino acids, particularly R, in their amino-terminal half, and by evenly spaced hydrophobic amino acids, particularly L, in their carboxy-terminal half. The basic regions are

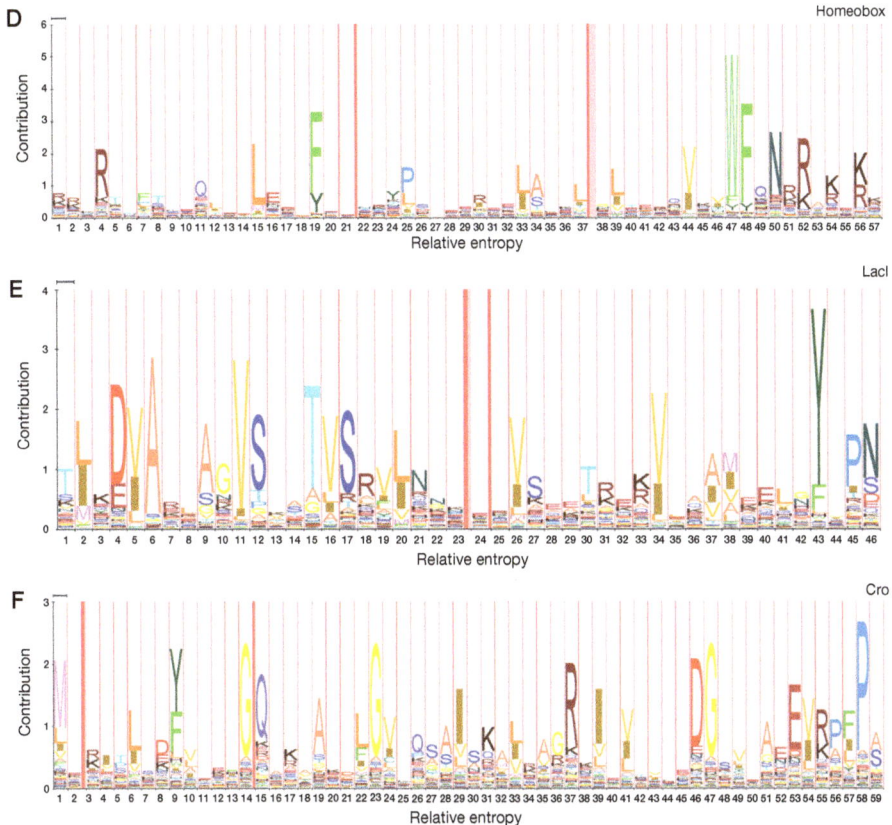

Figure 8-3. (*Continued.*) (*D*) Homeodomain transcription factors (homeobox, PF00046; 186 sequences). (*E*) lacI subfamily of helix-turn-helix (HTH) family of transcription factors (lacI, PF00356; 26 sequences). (*F*) cro subfamily of HTH family of transcription factors (cro, PF09048; six sequences).

involved in direct DNA contacts, whereas the hydrophobic "zipper" regions are involved in the pairing of helices to form dimers that bind DNA (see Fig. 3-5). One major difference between the families is the large insert region in the bHLH family (position 31 in the logo) that corresponds to different-size loop regions in different members of the family. The semiconserved H, E, and R at positions 5, 9, and 13, respectively, interact with the highly conserved CA of the E-box-binding site (CAnnTG) for bHLH family TFs. The most highly conserved N and R, at positions 13 and 21 of bZIP proteins, are also involved in direct DNA recognition.

Figure 8-3D–F shows logos for three different subfamilies of the helix-turn-helix (HTH) class of proteins: homeodomain, lacI, and cro, respectively. For each subfamily, semiconserved positions are among between the different subfamilies. Even though their DNA-binding domains are structurally similar, in detail their structures and modes of interacting with DNA are sufficiently different for them to have significantly different profile HMMs.

RECOGNITION MODELS

It has long been a goal to be able to predict the specificity of a TF based on its protein sequence. Structural information about which amino acids in a DNA-binding domain can interact directly with the DNA, coupled with a recognition code that defines how specific amino acids could select specific base pairs based on their individual properties (see Figs. 2-2B and 3-2B for diagrams of each base pair and amino acid), could provide predictions of preferred binding sites for any DNA-binding protein sequence. Such a code could also facilitate the design of new proteins with novel specificity. In 1976, before any DNA–protein complex structures were known, Seeman et al. (1976) pointed out that sequence specificity can be obtained by requiring two hydrogen-bond interactions in the major groove. These investigators further showed that Arg can make two hydrogen bonds specifically to a G-C base pair and Asn (or Gln) can make two hydrogen bonds specifically to an A-T base pair. A simple code for recognition could then be proposed, in which those amino acids, occurring in the appropriate positions on the DNA-binding surface of the protein, could specify any particular DNA sequence. Both of those interactions occur in the binding of the EcoRI restriction enzyme to its cut site (see Fig. 4-1), although several other contacts contribute to the very high specificity of that interaction, consistent with their further suggestion of additional types of interactions that could contribute to binding specificity. But when the first few complexes of a TF bound to DNA were determined by X-ray crystallography, it became clear that there was no simple and universal code for DNA–protein recognition (Matthews 1988). Although the interactions proposed by Seeman et al. (1976) do occur and are in fact quite common, there are many different ways that amino acids can interact with base pairs (see examples in Figs. 4-1– 4-4), and any particular amino acid may interact with either G-C or A-T base pairs, depending on the structural and sequence context.

Despite the lack of a simple, universal recognition code, considerable work has gone into determining recognition models, focusing mostly on developing predictive models for individual TF families. Because members of a TF family bind to DNA in very similar ways, there may be constraints on the interactions that define rules for interacting residues. The rules do not need to be absolute; even strong preferences for particular combinations of amino acids and base pairs can be useful for many purposes. Zinc finger proteins have been the most studied because of their modular binding domains. Most zinc finger proteins contain several fingers, and each one interacts with a short portion of the total binding site, typically three or four positions. Figure 4-2A shows the interactions between zif268 and its preferred binding site in which each of the three fingers interacts with four base pairs with a one-base overlap (the amino acid at position 2 interacts with the same base pair as the amino acid at position 6 of the previous finger) for a total binding site of 10 base pairs. This inter-action scheme is quite common among crystallized complexes and has come to be known as the canonical interaction, although other types of interactions are also observed (see other examples in Fig. 4-2). It was further discovered that fingers could be rearranged into different orders and binding-site preferences were largely pre-served, consistent with reordering of the protein. That meant that if one could deter-mine the appropriate zinc finger sequences to bind to any specific four-long DNA sequence, one could design zinc finger proteins composed of multiple fingers (typi-cally three or four) to bind with high specificity to a desired 10–13-long DNA sequence. Several groups of investigators began collecting such binding data, using complementary methods, and recognition codes for zinc finger proteins have been developed that are useful for designing proteins with desired specificity (see Box 4-1 and Fig. 8-4A) (Wolfe et al. 2000).

One method to determine the binding specificity of a zinc finger protein is to ran-domize a binding-site library and perform SELEX (see Chapter 5). In the early days, this was a low-throughput SELEX procedure where several rounds of selection were followed by sequencing a few of the selected sites, often resulting in a preferred sequence for that protein. This was usually done with a portion of the binding site being randomized—perhaps just the three or four bases that interact with a specific finger. By doing this for many different proteins, one could sense the preferences be-

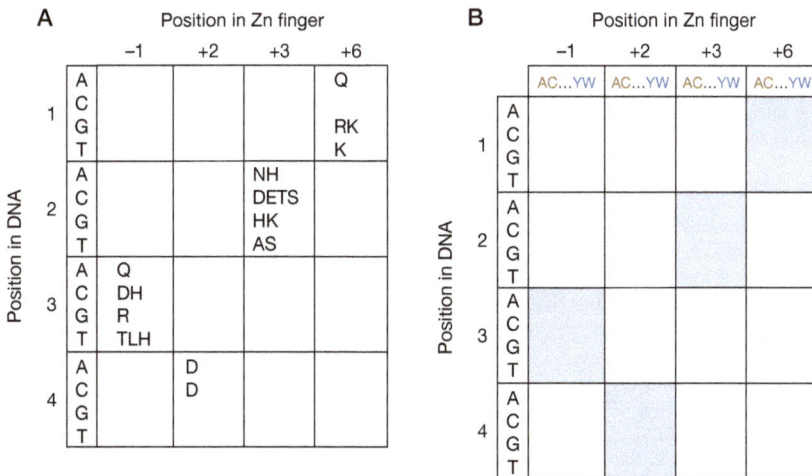

Figure 8-4. Recognition codes for zinc finger transcription factors. (A) Qualitative code for zinc finger proteins. This is equivalent to the code shown in Box 4-1 from Wolfe et al. (2000) but rearranged to emphasize the interacting positions between the protein and the binding site. (B) Qualitative code for zinc finger proteins, equivalent to A except that each shaded box would contain the 80 energy param-eters for all possible amino acids interacting with all possible base pairs. The 320 parameters (80 for each shaded box) are not shown but can be obtained, as determined by different approaches (see Joung et al. 2000; Benos et al. 2002). (Redrawn, with permission, from Benos et al. 2002; ©Wiley-Blackwell.)

tween specific amino acids at particular protein positions and positions in the binding sites could be seen. Recently, several high-throughput methods have been applied to this problem, such as protein-binding microarrays (PBMs), a high-throughput version of SELEX (SELEX-seq), and bacterial one-hybrid methods (B1H) described in Chapter 5. These approaches can provide extensive data about the binding specificity of a particular TF, including quantitative differences in binding affinities to a large number of sites.

The earliest complementary method is phage display (Fig. 8-5) (Smith 1985). In this method, the DNA sequence of the binding site is fixed and the protein is randomized to select for protein sequences that bind with high affinity to that DNA target sequence. For example, zinc finger proteins can be expressed as a fusion with a phage

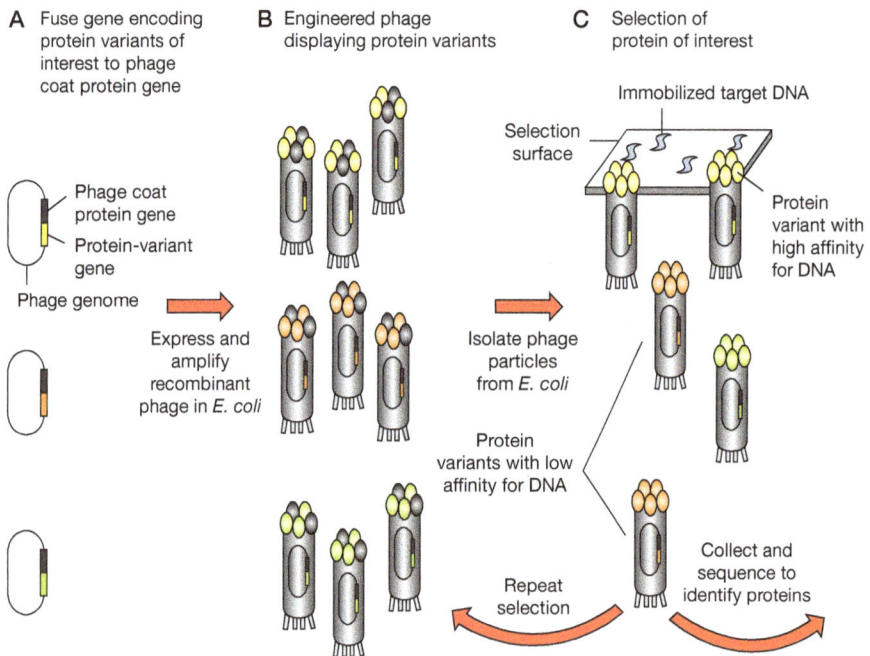

Figure 8-5. The phage genome (shown in *A*) contains a protein that is expressed on the tip of the phage (the black segment in *A* represents the gene and the black ball in *B* represents the protein on the tip of the phage). Different versions of the protein of interest are fused to that phage protein (differently colored segments in *A*) and they end up on the tip of the phage (colored balls in *B*). Only three examples are shown in the figure but, generally, many millions of different protein variants are generated. Each *E. coli* cell that is infected with a phage genome generates hundreds of copies of the phage with the protein of interest at its tips (*B*). Target DNA (blue waves) are attached to a surface (*C*). The phage particles pass over the target DNA, and those with high affinity will stick (yellow protein, in this case), whereas the other phage are washed away. The selected phage can be regrown in *E. coli* to amplify them. After a few rounds of selection, the resulting phage are sequenced to determine the proteins that have high affinity for the target DNA.

(bacterial virus) coat protein so that it is expressed on the surface of the phage where it can interact with DNA. The protein is randomized only in the positions that interact specifically with the DNA sequence, so that a moderate-size library can cover all possible protein sequences for those interacting positions. If the four critical amino acids from one finger are randomized, the total number of possible protein sequences is $20^4 = 160,000$. The phage are grown in *Escherichia coli*, and that number is easily accommodated; in fact, considerably larger libraries of $\sim 10^8$ different sequences can be infected into *E. coli* and used in the selections. The important aspect of this method is that DNA that codes for the protein is in the phage genome, so that sequencing the phage DNA will identify the selected protein sequences (protein sequencing is much less efficient than DNA sequencing). Furthermore, multiple rounds of selection can be performed by reinfecting *E. coli* with the selected phage from one round to generate a new library for the next round of selection. Each round selects for proteins with high affinity for the target DNA, and after several rounds, the highest-affinity proteins are readily identified. Recently, phage display has been replaced by bacterial one- and two-hybrid methods (B1H, B2H) (Joung et al. 2000). These are similar to the B1H method described in Chapter 5 for determining the specificity of a TF, but now the binding site is fixed and the sequence of the TF is randomized (for just the critical residues, as in the phage-display method). By selecting for fast-growing colonies, the collection of protein sequences that have high specificity (because the entire *E. coli* genome is competing for binding to the protein) for the target DNA sequence can be determined.

The early work, using SELEX and phage-display methods, generated hundreds of different combinations of high-affinity protein–DNA combinations from which recognition codes for zinc finger proteins were developed (see Box 4-1 and Fig. 8-4A). This is not the simple universal code originally sought because the preference depends on the position in the complex. A G at either position 1 or 3 of the binding site prefers an Arg at protein position 6 or −1, respectively, but G at position 2 of the binding site prefers His at position 3 of the protein. It is also degenerate in both directions. For example, G at binding-site position 2 prefers either His or Lys at protein position 3 (although His is selected somewhat more often), but that His is also preferred by A at position 2 of the binding site (somewhat less often than Asn). Given this code, a protein can be designed with multiple fingers to bind to almost any target sequence. When such designs are tested, they are often not as specific as one would hope because additional sequences are bound with similar affinities but can be good starting points for further selections to obtain proteins with high specificity for a particular target sequence.

One of the limitations of the recognition code of Figure 8-4A is that it is inherently qualitative. No attempt is made to predict the differences in binding energies of different DNA sequences to the same protein. Because zinc finger proteins, like all TFs, do not bind in an all-or-none fashion, it can be more useful to have a quantitative model, or probabilistic recognition code, that predicts the relative affinities to different sequences. A quantitative model can be defined that still imposes the same constraints on which amino acids interact with which positions in the binding sites, as shown in Figure 8-4B. The qualitative model of Figure 8-4A contains lists of preferred amino acids for each possible base pair at each position of the binding site, but in Figure 8-4B, those lists are replaced by binding-energy values for each possible amino acid interacting with each possible base pair; that is, there are 80 values within each square of the total matrix. Placing values only into the four shaded boxes enforces the canonical interaction model, where each of the amino acids at positions −1, 2, 3, and 6 interacts with a specific position in the binding site. The model can allow other interactions by simply including values for other squares in the matrix or even including additional positions. The main problem is that each new allowed interaction has a large number of parameters that must be estimated, requiring large increases in the sample size. Another limitation is that the overall model still requires independence; even if an amino acid can interact with more than one base pair, the total binding energy is the sum of its independent interactions.

Different quantitative models have been developed that are similar to what is shown in Figure 8-4B (Benos et al. 2002; Kaplan et al. 2005). These models each provide better predictions about the binding specificity of individual proteins than do qualitative models, but they also leave ample room for improvement. Other methods have been developed that use a variety of machine-learning approaches and they can improve the predictions even further. But so far, this remains a challenging problem without a very adequate solution. An alternative approach is to develop better molecular modeling software that can predict changes in binding energies for different protein and binding-site sequences based on biophysical parameters and molecular dynamic simulations, as is described briefly in Chapter 6. Some improvements in that approach have been made, and a few successful designs of new proteins with novel specificity have been generated, but this also remains an approach that is far from providing adequate predictions in general.

Chapter 7 shows how specificity of TFs could be determined from a variety of experimental data. Based on information about coregulated genes or from phylogenetic conservation, it is even possible to identify motifs when the TFs that bind to them are unknown. In this chapter, we show how TFs can be readily identified by their sequences. If the recognition modeling described above was successful, it would help to make the connections between the motifs and TFs that bind to them. The motifs for each TF could be predicted and the one that best matched the discovered motif could then be identified. Currently, recognition models only exist for a few of the most common TF families, such as zinc fingers and homeodomains, and even those are less accurate than one would like. As more extensive data accumulate, it may soon be possible to develop more accurate predictive models for a wider range of TF families, which would also help in the design of TFs to recognize desired motifs. Both of those abilities will aid in the goals of systems and synthetic biology, described in Chapter 9.

REFERENCES

Altschul SF, Gish W, Miller W, Myers EW, Lipman DJ. 1990. Basic local alignment search tool. *J Mol Biol* **215:** 403–410.

Benos PV, Lapedes AS, Stormo GD. 2002. Probabilistic code for DNA recognition by proteins of the EGR family. *J Mol Biol* **323:** 701–727.

Durbin R, Eddy SR, Krogh A, Mitchison G. 1998. Biological sequence analysis: Probabilistic models of proteins and nucleic acids. Cambridge University Press, New York.

Eddy SR. 1998. Profile hidden Markov models. *Bioinformatics* **14:** 755–763.

Henikoff S, Henikoff JG. 1992. Amino acid substitution matrices from protein blocks. *Proc Natl Acad Sci* **89:** 10915–10919.

Joung JK, Ramm EI, Pabo CO. 2000. Bacterial two-hybrid selection system for studying protein–DNA and protein–protein interactions. *Proc Natl Acad Sci* **97:** 7382–7387.

Kaplan T, Friedman N, Margalit H. 2005. Ab initio prediction of transcription factor targets using structural knowledge. *PLoS Comput Biol* **1:** e1.

Krogh A, Brown M, Mian IS, Sjölander K, Haussler D. 1994. Hidden Markov models in computational biology. Applications to protein modeling. *J Mol Biol* **235:** 1501–1531.

Matthews BW. 1988. Protein-DNA interaction. No code for recognition. *Nature* **335:** 294–295.

Seeman NC, Rosenberg JM, Rich A. 1976. Sequence-specific recognition of double helical nucleic acids by proteins. *Proc Natl Acad Sci* **73:** 804–808.

Smith GP. 1985. Filamentous fusion phage: Novel expression vectors that display cloned antigens on the virion surface. *Science* **228:** 1315–1317.

Smith TF, Waterman MS. 1981. Identification of common molecular subsequences. *J Mol Biol* **147:** 195–197.

Wolfe SA, Nekludova L, Pabo CO. 2000. DNA recognition by Cys2His2 zinc finger proteins. *Annu Rev Biophys Biomol Struct* **29:** 183–212.

ONLINE RESOURCES

InterPro (www.ebi.ac.uk/interpro) An integrated database supported by the European Bioinformatics Institute that links to many other databases with valuable information about protein sequences and structures.

Pfam (pfam.janellia.org, pfam.sanger.ac.uk, or pfam.sbc.su.se/) Pfam is a database of profile HMMs for many protein families. It also provides protein alignments and links to other resources, such as structural databases.

CHAPTER 9

Transcriptional Genomics

THIS FINAL CHAPTER RETURNS TO THE TOPIC OF GENE regulation mediated by protein–DNA interactions. Chapter 1 described the discovery, more than 50 years ago, of the *lac* regulatory system and how expression of the lactose-metabolizing gene *lacZ* was controlled by the Lac repressor binding to DNA to prevent transcription when no lactose was available. Chapter 1 also described other regulatory systems, such as the λ genetic switch and some novel characteristics of eukaryotic gene regulation that were among the early studies of protein–DNA interactions and the control of gene expression. The intervening chapters have all focused on different aspects of protein–DNA interactions, mostly as isolated systems studied in vitro. X-ray diffraction data led to the discovery of the structure of DNA, and protein structures have been determined by crystallography since the 1950s; however, it was only in the 1980s that the first structures of protein–DNA complexes were determined. The number of complexes with known structures has grown rapidly since then, and many structural principles related to binding affinity and specificity have been determined. Thermodynamic and kinetic studies of protein–DNA interactions began as soon as purified Lac repressor was available and have continued vigorously ever since on a wide range of transcription factors (TFs). Many technological advances have increased the sensitivity and throughput of data collection, allowing for much more sophisticated analyses of the energetic contributions to binding specificity. The development and continued improvement of computational modeling approaches to the study of structures and their dynamics and energetics has also led to new insights. Among the most significant technological advances was the invention of DNA sequencing methods in 1977, coupled with the ability to synthesize DNA of any desired sequence a few years later. These technologies have also advanced enormously. Complete genome sequences have been determined for a large collection of species, and it is now possible to determine the sequence of individual humans (or individuals of any species) cheaply and rapidly. DNA synthesis methods are also much faster, cheaper, and more reliable, allowing for many experiments with synthetic DNA that were not possible only a few years ago. Soon after DNA sequences became readily available, bioinformatics approaches began to analyze those sequences for important features, including the binding sites of TFs. Although there is still a lot to learn, the combined advances from many different approaches have led to enormous improvements in our understanding of DNA-binding protein families and how they interact with DNA to attain the specificity necessary for regulatory systems to function effectively.

Using the insights gained from the experimental approaches described in the preceding chapters, and especially the bioinformatics methods from Chapters 7 and 8, we return here to considering what happens inside cells—how protein–DNA interactions control gene expression. This is an enormous field of research, and this chapter serves as an introduction by describing a few specific examples in which the knowledge of protein–DNA interactions contributes to our understanding of important biological processes. An extensive reference list at the end of the chapter provides pointers to the vast literature of this topic.

The analysis of gene regulation can be studied at many different levels of detail and scale: from the interactions of specific factors that control the expression of a single gene to the entire regulatory network of a cell. This chapter is titled "Transcriptional Genomics," rather than "Regulatory Genomics," to emphasize the fact that we are only considering the regulation of gene expression at the level of transcription (and really just initiation of transcription), whereas a complete regulatory network for a cell must include posttranscriptional regulation as well. Multiple events affect gene expression after the initiation of transcription, from alternative splicing to regulation of translation and messenger RNA (mRNA) degradation. Proteins are also regulated by modifications and interactions with ligands or other proteins that affect their localization, function, and degradation. A comprehensive regulatory network for a cell would integrate all of those events and is beyond our current capabilities. However, transcription is a major contributor to gene expression; it is the most studied and best understood of the steps controlling gene expression, and identifying transcriptional regulatory networks can provide significant insight into cellular behavior and phenotype. We focus particularly on how information about, and models of, TF–DNA interactions contribute to our understanding of the control of gene expression in different cells and in different conditions, including the design of regulatory systems in synthetic biology applications.

GENE REGULATORY NETWORKS

The field of systems biology is devoted to understanding and modeling biological networks, particularly metabolic networks and gene regulatory networks (GRNs) and sometimes the combination of the two. GRNs are readily modeled as graphs, with nodes being individual genes and connections being directed edges from the regulatory gene to the regulated gene (Fig. 9-1). An edge can have a sign depending on

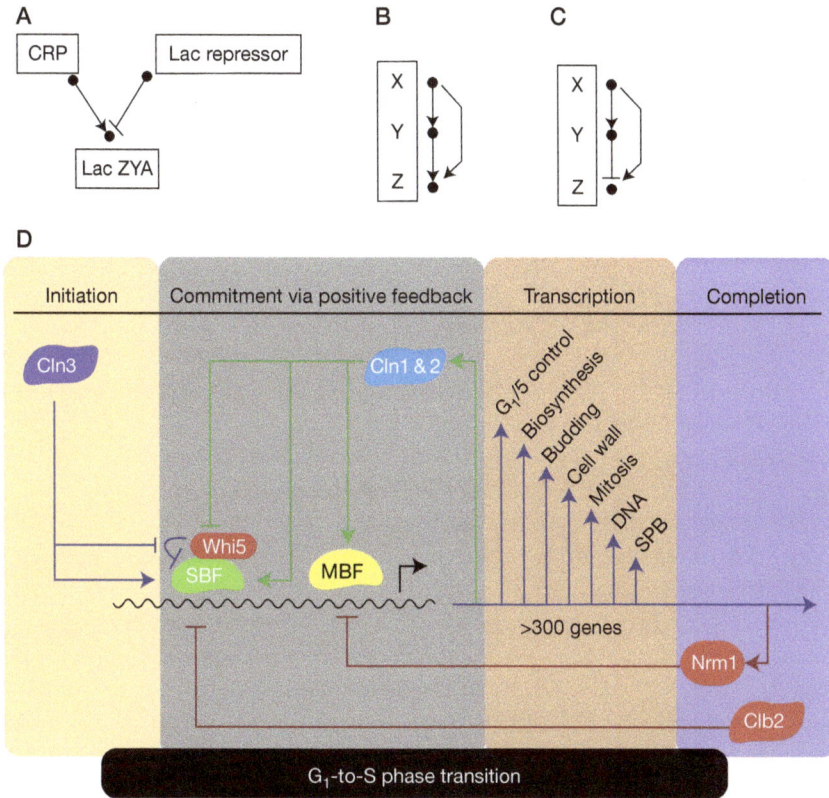

Figure 9-1. Basic elements of network graphs. (*A*) Regulation of the Lac ZYA operon by CRP (positive) and Lac repressor (negative). (*B*) Coherent feed-forward loop. (*C*) Incoherent feed-forward loop. (*D*) Regulatory network for cell-cycle control in yeast. Cln3 initiates the process by phosphorylating Whi5, which releases the inhibition of SBF. SBF then begins transcription of MBF, and they both activate many genes, including the genes needed for DNA replication. SBF is in a positive-feedback loop because transcription of Cln1 and Cln2 leads to further activation of SBF and MBF. After DNA replication is completed, Nrm1 and Clb2 reset the system to "off." (Redrawn, with permission, from Eser et al. 2011; *D*, © Elsevier.)

whether the regulatory gene activates or represses the regulated gene; this is typically indicated by an arrow for activation and a bar for repression. In some cases, a regulatory gene may have both functions, even on the same regulated gene, depending on the sites occupied or the cofactors engaged. One such example is the autoregulation of the λ repressor, which initially activates its own expression by cooperative binding to OR1 and OR2 and then represses its own expression at higher concentrations when OR3 becomes occupied (see Fig. 1-2). Although it is incomplete, the "wiring diagram" showing which proteins control the expression of other proteins and in which direction—activation or repression—can be very useful. Even a Boolean treatment of the network, in which each gene is either on or off, can lead to interesting dynamics in simulations where gene expression is followed through time. Logic functions can be applied, in which the state of the regulated gene depends on combinations of the states of the regulatory proteins (AND, OR, NOT, and more complex functions). Of course, real proteins are not simply expressed or not; they can have variable concentrations, and the kinetics of their production and degradation contribute to the network's complexity. Thus, rate parameters can be added to the edges of network graphs to make more realistic models that can be simulated with differential equations (Hasty et al. 2001). Small-molecule effectors that control whether specific factors are active or not, such as lactose for the Lac repressor and cAMP for the cAMP receptor protein (CRP) protein (see Chapter 1), can be added to the network. Other critical steps, such as protein modifications including phosphorylations that can activate or inactivate TFs, can also be included. A variety of computational approaches for modeling regulatory networks have been developed, including the development of languages such as the Systems Biology Markup Language (SBML), specifically for modeling, analysis, and simulations of biological networks of arbitrary complexity (see Resources at the end of this chapter).

Several results emerge from the analysis of real biological networks. One obvious characteristic is that, because typically only 5%–10% of genes are TFs, most TFs regulate multiple genes. Less obvious is that fact that most genes are regulated by multiple TFs and that such combinatorial regulation can lead to complex expression patterns. One feature is that certain "network motifs," which are specific patterns of connections, are very common (Alon 2007). Analyses of these recurring circuits show that they have desirable characteristics, such as fast response times to changing signals or stability in the presence of noise. For example, both positive and negative autoregulation, in which a protein controls its own expression, are quite common. The λ repressor and Cro proteins (see Chapter 1) are both autoregulatory, and the λ repressor is both positively and negatively autoregulated, which is a very efficient means for maintaining a constant concentration. Although not described in Chapter 1, the CRP protein is also negatively autoregulated, and it is estimated that more than half of the *Escherichia coli* TFs use autoregulation.

A more complex common network motif is the feed-forward loop (FFL), which involves three genes: X, Y, and Z. X regulates both Y and Z, and Y also regulates Z (Fig. 9-1B). Each of the three edges may be positive or negative, and thus there are eight possible types of FFLs. Furthermore, because Z is regulated by both X and Y, its output may depend on a logic function, such as AND or OR, or may be a simple additive response. Although there are eight possible FFLs, only two are quite common: One is the coherent FFL, in which all of the connections are positive, and another is an incoherent FFL, in which both of the edges from X are positive but Y negatively regulates Z (Alon 2007). These two types of FFLs have distinct beneficial characteristics, explaining why they are used so commonly. Of course, more complex network motifs, and combinations of motifs, also exist, and many are quite common in certain types of networks, such as developmental pathways where cells make decisions about the types of cell that they become. Another common occurrence in regulatory networks is positive feedback, which gives rise to a stable switch. Once a particular TF has been activated, its target genes provide further activation to maintain the "on" state of the TF. Such a system is used in both yeast and human cells to initiate DNA synthesis and the cell cycle (Fig. 9-1C) (Eser et al. 2011). Once the cell cycle has been initiated, it is vital that it be completed before the system is reset to the "off" state; such a positive feedback mechanism ensures that.

The above discussion about GRNs has made no mention of the genome sequence or the specificity of TFs. It is understood that TFs use sequence-specific binding to direct them to the genes that they regulate, but network analysis can be performed at a high level, where those details about exactly where, and to what sequence, the protein binds are not necessary. For example, using genetic and/or biochemical methods, one can determine that a particular TF is required for the proper expression of a certain gene and whether that influence is positive or negative, which provides a specifically labeled edge in the regulatory network graph. But additional information about the specific binding sites provides details that can be used in a variety of ways to increase our understanding of the regulatory network. In the following sections, we describe experimental methods for identifying the TFs that regulate specific genes, and/or bind to specific genomic regions, and the binding-site motifs that they recognize. We then provide a few examples of GRN analysis and the value of including information about the specificity of the TFs and their precise binding sites.

EXPERIMENTAL METHODS FOR IDENTIFYING
IN VIVO TF–DNA INTERACTIONS

Expression Analyses

Figure 7-5A showed a collection of *E. coli* promoter regions that were known to bind to the CRP protein and could be used by motif discovery algorithms to identify the CRP binding motif (such as a position weight matrix [PWM]). These data were obtained by a combination of genetic and biochemical experiments to identify genes whose expression depended on CRP; in fact, the correct binding sites were already known and that data set served as a benchmark to test the ability of algorithms to discover the motif (Stormo and Hartzell 1989; Lawrence and Reilly 1990). In the following years, new technologies, especially microarrays (Fig. 9-2A) (Schena et al. 1995), made it much easier to identify genes whose expression depended on particular TFs and even to identify genes that were coregulated presumably by having binding sites for some common, but perhaps unknown, TF. In microarray experiments, the amount of mRNA corresponding to each gene is measured and can be compared

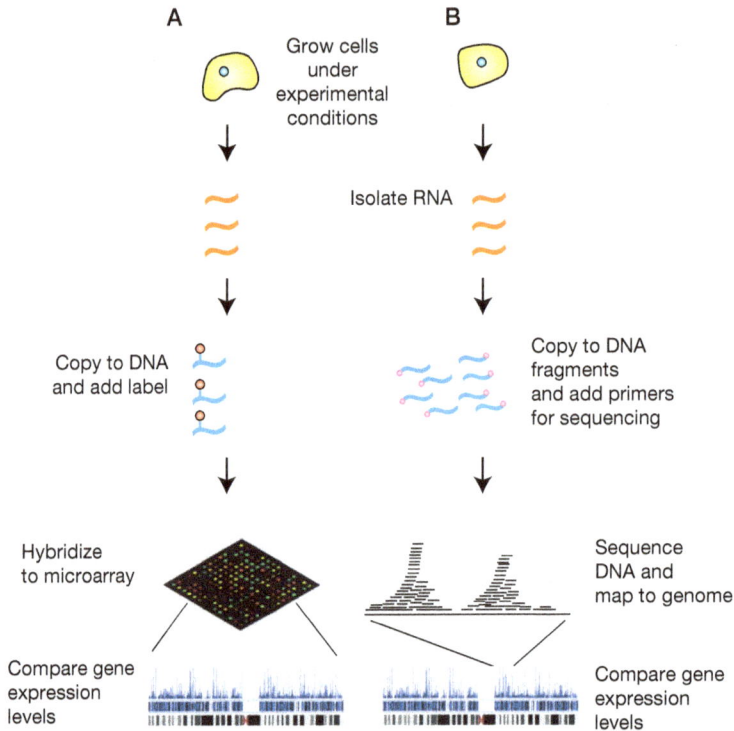

Figure 9-2. Assays for mRNA expression. (*A*) Microarray analysis of gene expression. mRNA from cells is copied into DNA (cDNA), amplified, fluorescently labeled, and hybridized to microarrays containing DNA for each gene. The fluorescent signal at each spot is proportional to the mRNA expression level of the gene corresponding to the spot. (*B*) RNA-seq analysis of gene expression. mRNA from cells is copied into DNA, amplified, and sequenced. Mapping reads back to the genome quantifies the expression level of each gene. (Modified in part from Kim and Park 2011, with permission from John Wiley & Sons Inc.)

among different types of cells and cells grown in different environments and to follow changes in expression over time. Of particular value are expression profiles for cells with and without a specific TF; for example, when it has been made nonfunctional by mutation or inactivated by some experimental manipulation. Those comparisons can readily identify genes whose expression depends on that TF and may be regulated by the TF (being aware that the dependence may be indirect). That information provides connections in the regulatory network and can also be used by motif discovery algorithms to infer the binding motif of the TF. By comparing gene expression profiles under different conditions, such as their response over time to a specific stimulus, one can identify sets of genes with similar expression patterns that may, therefore, be coregulated by a common TF or set of TFs. In this case, the TFs may not be known, but motifs can still be found and the important regulatory sites identified

(Roth et al. 1998). The advent of new high-throughput sequencing technologies has fostered the use of mRNA sequencing (RNA-seq) instead of microarrays to measure the relative expression of each gene (Fig. 9-2B) (Wang et al. 2009). Information obtained from either approach can be very useful in identifying genes with interesting expression patterns, making network connections between TFs and the genes that they regulate, and in many cases identifying the binding sites and motifs for TFs of interest.

An alternative approach uses reporter assays to make network connections and identify regulatory binding sites. The reporter is a gene whose expression is easy to measure, including a fluorescent protein such as green fluorescent protein (GFP), or whose expression can be subject to selection so that cells only grow when the gene is expressed at a reasonably high level. Reporter assays have been done in many different ways and for a variety of purposes. The following are a few recent examples that use high-throughput reporter assays to help make connections in regulatory networks.

A yeast one-hybrid system (similar to the bacterial one-hybrid system described in Chapter 5) has been used to identify *Caenorhabditis elegans* TFs that bind to regulatory regions from throughout the *C. elegans* genome (Walhout 2006). A regulatory region is cloned upstream of a reporter gene in the yeast genome and then a library of TFs is transformed into the cells. Sequencing from the cells that are positive for reporter expression can identify all of the interacting TFs in parallel. If multiple different regulatory regions interact with the same TF, motif discovery algorithms can be used to identify its binding motif.

For sea urchin genes whose expression is of interest but whose regulatory regions were unknown, candidate regulatory regions were rapidly screened to identify those that activate gene expression (Nam et al. 2010). Each candidate regulatory region was cloned with a specific identifying sequence (a "barcode") within the reporter gene transcript. The entire collection of candidate regulatory regions was injected into sea urchin eggs, and the expression driven by each region could be measured in parallel by mass sequencing of mRNA obtained from the eggs at various time points. This information was sufficient to detect the activating regions, which were ~20% of those tested, determine the times when each was active over a 48-h time span, and measure the relative activity of each region.

Very-high-resolution determination of the sequence contributions from enhancers has been determined using synthetic DNA to drive expression of a reporter gene containing a barcode for each variant. In one study (Melnikov et al. 2012), the synthetic DNA contained all possible single-base mutations of the enhancer region and a large collection of double mutations, a total of >27,000 enhancer variants. By simply sequencing the barcode regions from the mRNA isolated from cells, the investigators could determine the relative expression of each variant in parallel. The results show, at single-base resolution, the locations of the binding sites for TFs that control expression, sites that match the motifs for the known regulatory TFs for these enhancers. They could further develop a quantitative model for the effects of different sequence variants and design-modified enhancers with optimized responses. In another study (Patwardhan et al. 2012), a low frequency of random mutations was synthesized into three enhancers, and >100,000 variants of each were assayed for activity by sequencing the reporter gene containing a barcode sequence. Every position was mutated multiple times, and regression analysis allowed the authors to determine the contribution to enhancer activity of all possible single-base variants.

Binding Location Analyses

One can also determine directly the locations within the genome that TFs bind using chromatin immunoprecipitation (ChIP) methods (Fig. 9-3). Live cells are treated with formaldehyde, which cross-links TFs to the DNA at their binding sites. The DNA is fragmented, and the regions bound to the TF are isolated using antibodies to the TF (or to peptide "tags" that have been genetically added to the TF). The cross-links can be reversed, freeing up the isolated DNA that can then be labeled and applied to microarrays (ChIP-chip; Fig. 9-3A), or sequenced (ChIP-seq; Fig. 9-3B). Using microarrays that contain all of the intergenic regions from yeast, the binding-site locations for nearly all of the TFs were identified, many of them under more than one condition (Harbison et al. 2004). The resolution of the binding sites was typically a few hundred base pairs, but by using motif discovery algorithms (see Chapter 7), binding-site motifs for most TFs were identified. The ChIP-seq approach generates similar data to ChIP-chip but generally at higher resolution (Johnson et al. 2007). In this case, the data can come from anywhere in the genome, so one does not have to know in advance which regions to assay. For species with diploid genomes (two copies of each chromosome, one from each parent), if there are differences between the two chromosome sequences, ChIP-seq can even determine whether the TF binds preferentially to one of the two copies of a particular binding site (Ni et al. 2012). This can be very useful when only one copy of the gene is expressed (or one copy is expressed much more than the other), which can occur due to differences in chromatin between the two genes (epigenetic differences) or because variations between regulatory regions of the two genes influence the binding of TFs and regulation of expression.

Some other ways of determining the binding locations of TFs in vivo do not require cross-linking of the TF to DNA. The DamID method fuses the *E. coli* DNA adenine methyltransferase (Dam) onto a TF (or other protein associated with chromatin), which then methylates the A of the sequence GATC that occurs near the binding site of the TF (van Steensel and Henikoff 2000). Most species do not have enzymes that methylate GATC sites, so cutting genomic DNA with the restriction enzyme DpnI, which cuts only at methylated GATC sites, can be used to identify locations of binding sites for the TF. A similar method is referred to as "calling cards," in which a modified TF can

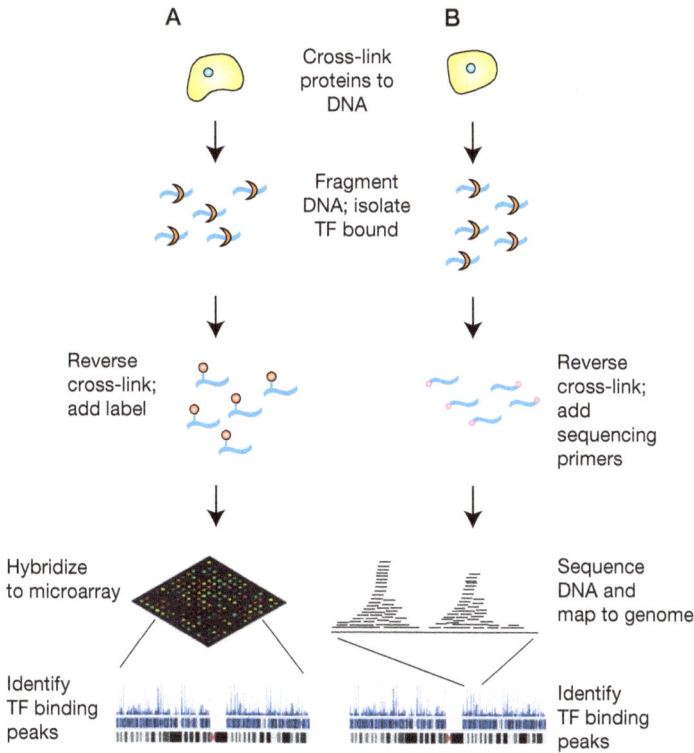

Figure 9-3. Localization of TF binding sites. (A) Chromatin immunoprecipitation (ChIP) followed by microarray (chip) analysis (ChIP-chip). Proteins are cross-linked to DNA with formaldehyde and the DNA is isolated and fragmented. DNA bound to a specific TF is precipitated with an antibody to that TF (often, the antibody is to a peptide tag added to the TF protein). The precipitated DNA is amplified, fluorescently labeled, and then hybridized to a microarray (chip) covering the regions of interest in the genome. Quantification of the fluorescent signal identifies the binding sites in the genome. (B) ChIP is performed as in A, but the isolated DNA is amplified and sequenced. Mapping the reads back to the genome provides a quantitative measure of TF binding throughout the genome. (Modified in part from Kim and Park 2011, with permission from John Wiley & Sons Inc.)

induce the insertion of DNA into the genome near to its binding sites (Wang et al. 2012a). Unique calling-card sequences can be associated with different TFs, and their locations in the genome can be identified in parallel by sequencing outward from the calling card into the surrounding genomic DNA (Wang et al. 2011). An advantage of calling cards is that, unlike the methylation of GATC sites, the DNA insertions are permanent modifications to the genome that will be passed on to the progeny of the cell where they occur (as long as the DNA insertion does not disrupt an essential function, which is generally the case). This means that the insertions can be used to track the locations of TF binding sites associated with cell differentiation pathways even if the critical binding events occur many generations before the final differentiation state.

DNA Accessibility Analyses

Most of the DNA in eukaryotic genomes is contained in nucleosomes, which are complexes of histone proteins with ~150 bp of DNA wrapped around them (see Fig. 1-3 and 9-5 below). Regulatory regions, including promoters, enhancers, and other types, are generally segments of the DNA in which nucleosomes have been replaced by complexes of regulatory proteins. DNA within nucleosomes is resistant to cutting by DNase I, an enzyme that cuts any DNA sequence. Regulatory regions, although bound by some proteins, contain much more accessible DNA and are therefore DNase I hypersensitive sites (DHS). The DNA from nuclei that have been treated with DNase I can be subjected to high-throughput sequencing, and the ends of those sequences, which are the sites cut by the enzyme, are highly enriched in regulatory segments of the genome (Hesselberth et al. 2009). An alternative approach for identifying nucleosome-free regions of the genome is formaldehyde-assisted isolation of regulatory elements (FAIRE) in which formaldehyde cross-linked DNA is sonicated to break it into small segments and then the non-nucleosome-associated DNA is isolated, labeled, and hybridized to microarrays (similar to the ChIP-chip experiments described above) (Giresi et al. 2007). The regions identified by FAIRE and DHS are generally in very close agreement and both can be used to identify regulatory regions in a genome-wide assay.

FAIRE and DHS analysis have both advantages and disadvantages compared to ChIP-chip and ChIP-seq methods. The main advantage is that, in a single assay, one identifies all of the regulatory sites in the entire genome. The disadvantage is that the factors that bind to those regions are unknown, whereas with ChIP methods, antibodies to specific TFs (or other proteins) are used to isolate the bound DNA segments. However, by using catalogs of motifs, such as PWMs, for the TFs encoded in the genome, one can often identify many of the TFs that are likely to bind to specific regulatory regions (Pique-Regi et al. 2011). This is especially true if one knows the subset of TFs expressed in the specific cells that are assayed. Currently, this approach is limited because motifs are known for only a fraction of the TFs in any higher eukaryotic species, but as more information is gained about the specificities of TFs, it will be a very efficient method for making connections between TFs and the genes that they regulate on a genome-wide scale for any particular cell type and genetic background (see the ENCODE section below).

STUDYING GRNs

Knowing all of the TFs for an organism and the specificity of each would allow for the prediction of a complete transcriptional network of the TFs and their regulated, or target, genes. Of course, just knowing the TFs and their targets would be a very incomplete model of the regulatory network of the cell for a variety of reasons.

This not only leaves out all of the nontranscriptional regulation that occurs but also misses information about which signals the TFs respond to, whether they affect their targets positively or negatively, and which TFs function cooperatively and how that affects their activity. In eukaryotic cells, this also omits the role of DNA accessibility; the majority of predicted binding sites that contain high-affinity sequences are not bound by the TF because they are not accessible, or in many cases, they are bound but apparently do not contribute to gene regulation. Accessibility depends in part on the epigenetic state of DNA segments, which includes modifications to the DNA as well as to the histones, and the underlying causes of those states are not well understood. Despite their limitations, transcriptional networks are extremely valuable in understanding cellular phenotypes and responses to perturbations, and much progress has been made in recent years to increase their completeness. The following sections describe a few examples from a variety of organisms of our current knowledge of GRNs. Because of their size and complexity, and the need to understand their dynamics, computational approaches are integral to modeling regulatory networks. The development of algorithms and programs is a very active area of research. A few recent reviews on this topic from several different perspectives have appeared (Babu 2008; Hecker et al. 2009; Janky et al. 2009; Ghosh et al. 2011; Kim and Park 2011; Aerts 2012).

Bacteria

E. coli is by the far the most-studied bacteria, and its regulatory network is the best characterized. K12, the best-studied *E. coli* strain, has a genome of ~4.6 million base pairs, but other strains can be >10% larger. K12 has ~4300 genes, of which >5% are known or predicted to be sequence-specific TFs. There are databases, such as RegulonDB and EcoCyc (Keseler et al. 2011; Salgado et al. 2012), that contain information about *E. coli* TFs, including their specificities (as PWMs) and the genes that they regulate. They track experimentally identified binding sites and can make network graphs based on those interactions. Using PWMs, binding sites can be predicted throughout the genome, expanding the regulatory network to include putative and potential interactions. Many different strains of *E. coli* have also been sequenced, and differences in regulation can sometimes be predicted based on sequence variations. Combining information from metabolic and regulatory networks can provide a more integrated view of cellular behavior and response to environmental changes (Herrgard et al. 2004). A few other bacterial species have also been studied extensively, and similar databases exist for their regulatory interactions (for review, see Janky et al. 2009). Notable are species from the *Salmonella* genus (closely related to *E. coli*) and the *Bacillus* genus (distantly related to *E. coli*). One fact that stands out for TFs with very similar DNA-binding domains, especially if the contacting residues are identical, is that the specificities are conserved, even though the genes that are regulated by the TFs may be different. The evidence that regulatory networks can change rapidly through evolution comes from comparisons among many different species (Janga and Collado-Vides 2007; Perez and Groisman 2009). Although specificities of TFs are determined by their protein sequences, their binding sites in the genome are easily changed and lead to "rewiring" of the regulatory circuitry. One class of sequence-specific binding proteins not described in the previous chapters is the bacterial σ factors (Gruber and Gross 2003). These are an interchangeable subunit of the RNA polymerase complex that synthesizes RNA from the DNA template. Bacteria typically have several different σ factors, each conferring specificity for a subset of bacterial promoters. In *Bacillus subtilis* and related species, the generation of spores is controlled by alternative σ factors, some of which are localized to the spore-forming cell and not in the "mother cell" that gives rise to the spore (Sierro et al. 2008; de Hoon et al. 2010).

In addition to the few bacterial species that have been studied extensively for many decades, new technologies and initiatives have greatly expanded the repertoire of sequenced bacterial genomes. There are currently well more than 1000 bacterial species with completely sequenced genomes and even more with partially sequenced genomes. For example, the Human Microbiome Project (see Resource list), funded by the National Institutes of Health, aims to identify the thousands of different species of bacteria that inhabit different locations on and in the human body. The repertoire of species can be identified by amplifying and sequencing the ribosomal RNA genes. Rather than isolating and sequencing individual species, many of which are not culturable by current methods, a "metagenomic" sequencing approach is used, where short sequencing reads are obtained from the mixture of all the genomes. The most abundant species generate enough reads that they can be completely assembled, but rare species generally have just a portion of their genomes represented. This approach has not only been applied to the bacteria associated with humans, but in fact, many different environmental samples have been analyzed by metagenomic sequencing. For the vast majority of bacterial genomes now available, both complete and partial, essentially nothing is known about their biology from experiments. Nevertheless, many gene functions can be inferred by analyzing the sequences and comparing them to existing sequences of known function. MicrobesOnline (Dehal et al. 2010) is a database containing genome sequence information and many other types of data for several thousand species including bacteria, archaea, and some eukaryotes. This includes TFs, and it is possible, at least in some cases, to also infer their specificity and to make inferences about their target genes. By comparing the promoters of operons for homologous genes in different species and by phylogenetic footprinting (see Chapter 7), in which the conserved segments of the promoters are identified, it is possible to infer the regulatory sites of those operons. The Regtransbase (Kazakov et al. 2007) database catalogs inferred TFs and binding sites across species based on direct experimental data as well as such bioinformatics approaches. Because TFs with conserved DNA-binding domains have the same specificity, and because many bacterial TFs are autoregulatory, it is also possible to apply motif discovery algorithms (see Chapter 7) to the promoters of those TFs to infer their binding sites (Sahota and Stormo 2010).

Synthetic Biology

The field of synthetic biology creates organisms with new capabilities by modifying or making additions to their genomes. The scale varies from the highly ambitious goal of creating an entire organism from synthetic DNA to more modest goals of adding substantial new genetic components to an organism. The latter approach is an extension of traditional genetic engineering, in which instead of just expressing a single gene, an entire metabolic pathway, or combinations of pathways, is created to synthesize desired products. Usually, the genes cannot be simply cloned from their natural species but instead must be optimized for expression in the new host, which is often *E. coli*. The new genes can also be synthetic genes that do not exist in nature, such as modified enzymes to use new substrates or generate new products. In each case, the expression of the genes requires control that is usually obtained from a catalog of existing promoters and TFs with their regulatory sites. One such catalog is the "registry of standard biological parts" (see Online Resources), a collection of "biobricks," components that can be combined to build synthetic biological systems. This catalog includes promoters for many different RNA polymerases and TFs and their binding sites for controlling expression by various effectors. And of course, those parts can be modified to optimize their activity for desired responses. The International Genetically Engineered Machine (iGEM) Foundation (see Online Resources) promotes education in synthetic biology by sponsoring annual competitions in which teams design and build bacteria to perform diverse tasks.

In addition to using synthetic biology to generate useful products, it is also interesting to make and study regulatory networks for their intrinsic characteristics and to test the ability to design networks with specific behaviors. One can design regulatory networks with particular properties, just as one would do for an electrical circuit, but the implementation of them in vivo into functional modules with the desired behavior is not always easy and often requires care in the choice of specific parameters, such as protein–DNA interaction affinities. Two examples are shown in Figure 9-4: One is a simple toggle switch that can be set to either state with an external signal, and the other is an oscillator with a period longer than the cell cycle, so that newly divided cells "remember" where they are in the oscillatory period. Several good reviews on synthetic biology describe those examples and many others and lay out new directions and applications for the field (Kaern et al. 2003; Guido et al. 2006; Drubin et al. 2007; Mukherji and van Oudenaarden 2009; Khalil and Collins 2010). Current DNA synthesis methods as well as computational modeling approaches create an enormous potential for synthetic biology, and some of the foundational technologies rely on the design of regulatory networks including, but not limited to, protein–DNA interactions.

A

Electronics	Biology	Synthetic biology
Reset-set latch	Bacteriophage λ lysis/lysogeny switch	Genetic toggle switch

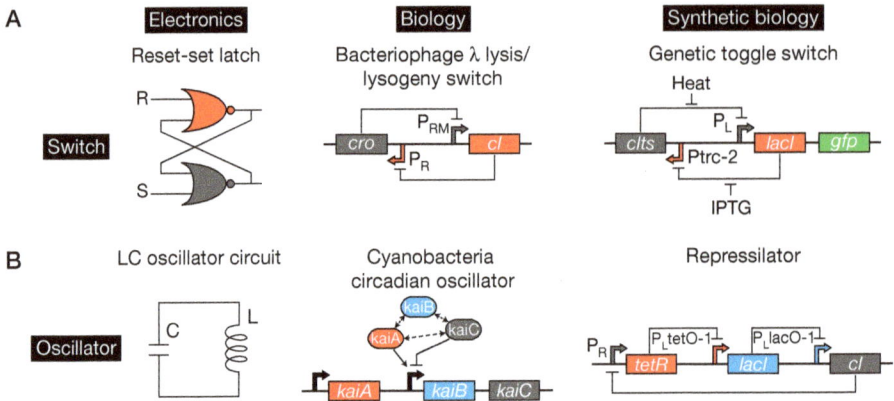

Figure 9-4. Bacterial synthetic gene networks. Diagrams of networks for a switch (A) and an oscillator (B). (*Left column*) Design for an electronic component, (*middle column*) a natural biological example, (*right column*) a synthetic biological example. (Reproduced, with permission, from Khalil et al. 2010; switch design reproduced, with permission, from Gardner et al. 2000; oscillator from data used, with permission, from Elowitz and Leibler 2000; © MacMillan Publishers Ltd.)

Yeast

The yeast *Saccharomyces cerevisiae* is by far the best-studied eukaryote for transcription. In fact, more is known about its regulatory network than for *E. coli*. Its genome is ~12 Mb, and there are ~6000 genes. Only ~200 of its proteins are known or predicted to be sequence-specific TFs, less than the 5% seen in most other organisms. Not only is there a long history of studying TF binding and gene regulation using genetics and biochemistry in yeast, but it is this organism for which ChIP-chip was originally developed and optimized (Lee et al. 2002; Harbison et al. 2004). In these studies, the binding locations for most yeast TFs were determined, many of them under multiple growth conditions. Binding-site motifs could be determined from that data for most of the TFs as could a preliminary network connecting TFs to their target genes. Not all TFs are active under the conditions tested thus far, and as a result there are a few TFs with still very limited data; however, there are now PWMs for the majority of these TFs. Several rounds of developing ever more complete regulatory networks for yeast and how they vary with the growth condition have occurred, and quite complete regulatory networks are known for some specific cellular processes, such as the cell cycle. The *Saccharomyces* Genome database (SGD; see Online Resources) contains extensive information about the genome and the genes that it encodes, and there are several databases specifically about yeast TFs and their specificities (see Online Resources). Several other yeast species have now been sequenced and allow one to observe evolutionary changes in regulatory sequences, in which binding sites are lost and gained, and to both model and measure the corresponding differences in gene regulation. Yeast is also used extensively in synthetic biology because many proteins and pathways can be synthesized more effectively, along with necessary posttranslational modifications, in eukaryotic cells than in bacteria. An interesting example is the generation of yeast strains to make the antimalarial drug precursor artemisinic acid (Ro et al. 2006). This not only required the expression of some plant enzymes in yeast but also the adjustment of the levels of several intrinsic yeast genes by modifying their regulatory sites. A collection of zinc finger TFs that recognize nonoverlapping sets of binding sites has been developed and can be used to control distinct regulatory networks, opening up further applications of synthetic biology in yeast (Khalil et al. 2012).

The ENCODE Project

Much less is known about the regulatory networks of higher, multicellular eukaryotes than of bacteria and yeast. The problem is much more complicated for several reasons. One is that the genomes are larger: $\sim$100 Mb for model organisms such as worms and flies, but >1 Gb (one gigabase = 10^9 base pairs) for most species of interest (the human genome is $\sim$3 Gb in total length). Much larger sizes mean that there are many more potential regulatory regions and TF binding sites, and regulatory regions may be located at considerable distances from the genes that they regulate. Making the connections between TF binding sites and target genes is challenging. There are also more genes, and more TFs, than in the simpler organisms. For example, the human genome is predicted to have >20,000 protein-coding genes, most of them with multiple isoforms (alternatively spliced mRNAs) and nearly that many noncoding RNA genes, all of which are transcriptionally regulated. There are predicted to be >1400 sequence-specific TFs based on sequence analysis methods (see Chapter 8). Furthermore, in different cell types of multicellular organisms, the regulatory sites for each gene may be different, and individual TFs may regulate different subsets of genes. In fact, one of the most interesting and challenging problems in biology is how development occurs—how a single cell gives rise to all of the various cell types, in the appropriate numbers and positions, needed for a fully functional organism. Much of the information is encoded in the transcriptional regulatory networks that allow for differential gene expression over time and space during development.

The ENCODE (Encyclopedia of DNA Elements) project is an NIH-funded consortium of more than 30 research groups that is using a variety of techniques to uncover the functional elements in the human genome. A pilot project was initiated in 2003 to analyze in depth 1% of the human genome; these segments were chosen for a diversity of functions, including many without any known or predicted functions. At the time, the best available methods for detailed analysis were based on microarrays, which were used to measure both RNA expression and protein binding (ChIP-chip). It is easy to target particular genomic regions with microarrays because those are the only probes included on the arrays. But with the advent of sequencing-based methods, such as RNA-seq and ChIP-seq, one automatically interrogates the entire genome by mapping the sequences produced onto the genome sequence. On the basis of the success of the pilot project in identifying new functional regions across the 1% of the genome, in 2007 it was expanded in two ways. First, a subproject called modEN-CODE added the same types of analyses on two well-studied model organisms, *C. elegans* and *Drosophila* (worms and flies). Both species have been used for a variety of biological studies for many years; they are well characterized genetically, and many molecular biology approaches can be conveniently applied. The genomes of each have been sequenced, as have genomes from several closely related species. The results added considerably to what is known about the functional features of the genomes, including the binding sites for many TFs. The second expansion of ENCODE was to include the entire human genome in the analysis and to consider a much larger collection of cell types. The number of assays was also increased to assess more types of functional domains with a large repertoire of techniques. In September 2012, the ENCODE consortium published more than 30 papers simultaneously in several different journals (see Online Resources). Figure 9-5 provides an overview of the types of experiments that were performed. Assays included measurements of RNA expression, including direct measurements of the 5′ ends of RNAs, and ChIP-seq experiments for 119 transcription-related proteins, including 87 sequence-specific TFs, as well as specific histone modifications that are associated with different chromatin classes and functions. Identification of accessible regions of DNA was obtained by both DNase I hypersensitivity and FAIRE analysis. Also assayed were modified bases in DNA and the overall three-dimensional (3D) chromosome architecture by determining those segments of the genome that occur in close proximity to one another. There were 147 different cell types included in the analysis, although only a subset of analyses were performed for most cell types. Comparisons of the same data types among different cell types allowed for the identification of cell-specific gene expression patterns and TF binding locations that in many cases could be used to infer the specific regulatory regions controlling the expression of specific genes. In total, 1640 data sets were analyzed in these papers, all of which are available online from the ENCODE project (see Online Resources). In the following paragraphs, we describe the results from a few of the papers that are specifically related to TF–DNA interactions.

ChIP-seq Experiments

More than 450 ChIP-seq experiments were performed on a set of 119 transcription-related proteins, including 87 sequence-specific TFs, in 72 different cell lines (Gerstein et al. 2012; The ENCODE Project Consortium 2012; Wang et al. 2012b) and results are available from the ENCODE web site as well as factorbook.org (see Online Resources). Obviously, not all TFs were assayed in all cell lines, and this is a small

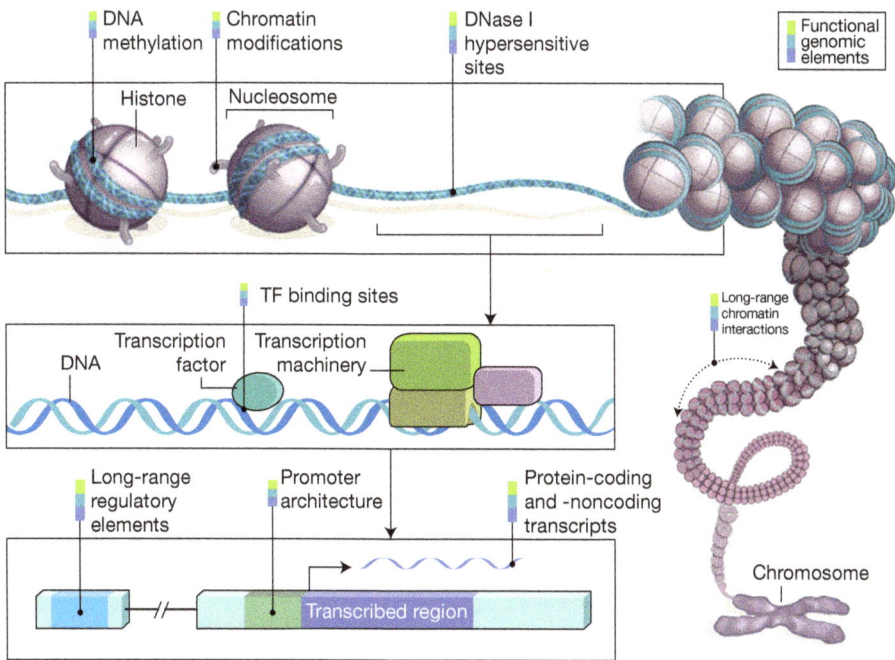

Figure 9-5. Overview of the type of information collected in the ENCODE project, including RNA sequencing to obtain the transcribed regions and both coding and noncoding RNA. The start sites of transcription are identified, including many alternative promoters, and many alternative splice junctions are found. ChIP-seq is used to obtain TF binding sites, locations of promoters, and histone modifications. Sites of methylated cytosines are identified. DNA-accessible regions are identified by DNase I hypersensitivity and by FAIRE analysis. Long-range chromatin loops are identified by capturing and sequencing chromatin segments in close proximity in 3D. (Reproduced, with permission, from Ecker et al. 2012; © MacMillan Publisher Ltd.)

fraction of the estimated 1400 TFs in the human genome; however, it represents a large increase over previous information about TF binding sites and regulatory regions across the entire genome. More than 7 million individual peaks were identified, covering more than 600,000 distinct regions and >230 Mb (8.1%) of the genome. PWMs could be found from peaks for 86 of the 87 TFs, most of which matched the known PWMs for those proteins but a few of which were novel. In most cases, there were also significant secondary motifs that may correspond to TFs that bind close to, and perhaps interact with, those used for the ChIP. In fact, the ChIP peaks for many TFs overlapped significantly, suggesting interacting pairs. Cell-specific differences occurred, as expected, in which the same TF binds to different locations in different cell types, and the same regions are bound by different TFs in different cells. Approximately 40% of the regions were within 2.5 kb of transcription start sites. The TFs and their target genes, using promoter-proximal sites, can be put into a network. Of course, some TFs regulate other TFs, and a hierarchy can be made in the regulatory network. In addition, it is found that FFLs (Fig. 9-1) are as common here as they are in bacteria and yeast.

Using ChIP-seq peaks and PWMs for six well-characterized TFs, Whiteld et al. (2012) predicted the precise binding sites in 455 promoter regions. Using reporter assays, the authors tested each region for function, along with mutated versions of each, in four different cell lines. Approximately 70% of the sites were active (in which active is defined as a significant difference in expression between the wild type and mutant sequence) in at least one cell type, and ~15% of the sites were active in all cell types. The difference could be either increased or decreased reporter gene expression, indicating either activation or repression. Given that each cell type had sites that were active only in that cell type, and that all of these sites were obtained from ChIP peaks for the associated TF, it is likely that they would all be active in some appropriate cell type. This shows that ChIP peaks, which are typically a few hundred base pairs long, combined with PWMs can locate the precise binding sites for a TF quite reliably. Furthermore, sites for the same factor may be activating in some sequences and repressing in others, although the majority of the functional sites were activating.

DNase I Hypersensitivity Experiments

DHSs were determined for 125 different cell types (Thurman et al. 2012). A total of nearly 2.9 million DHSs were obtained, with an average of ~200,000 per cell type. This represents ~1% of the genome in each cell type and a total of ~15%. Each additional cell type has overlap with existing cells but also adds new DHSs that are unique, with a plateau estimated at perhaps 40% of the genome (Stamatoyannopoulos 2012). Nearly all of the ChIP-seq peaks that were identified fall within DHSs, but a few were in non-DHSs, suggesting that they bind to nucleosomal DNA. DHS analysis was also performed at much higher read density on 41 cell types (Neph et al. 2012b). At that high density, DHSs can be resolved into footprints for individual TFs, often with several DHS footprints occurring within a particular DHS region. A total of 8.4 million distinct DHS footprints was observed in all cell types, with most occurring in multiple cell types but some unique to an individual cell type. A distinctive 50-bp pattern overlapped transcription start sites (TSSs) and presumably represents the footprint of the RNA polymerase complex bound to the promoter.

As described previously, an advantage of DHS mapping is that all of the proteins bound to DNA can be observed from a single experiment (except for those bound to nucleosomes), but the disadvantage is that the proteins giving rise to a particular footprint are unknown. Given a catalog of known motifs, one can compare these to the DHSs and make putative links between specific TFs and binding locations in the genome. Nearly all of the motifs from several catalogs of human PWMs (see Online Resources) were found to have matches in the observed footprints. These investigators also applied motif discovery algorithms (see Chapter 7) to the set of foot-prints and found 683 distinct motifs, 394 of which matched to the known catalogs and 289 were novel. This set of motifs is undoubtedly incomplete for several reasons, but it provides an unbiased collection of likely regulatory motifs and their individual binding sites, across the entire genome. Although there are predicted to be >1400 sequence-specific TFs in the human genome, the number of distinct motifs will be less because several TFs have motifs that are nearly identical and will be indistinguish-able without more sensitive analyses. The actual number of expected, distinct motifs is unknown, and the 683 discovered may represent a large majority of the complete set. However, any TFs that are not expressed in the set of cell types analyzed will not be included, and TFs with only a few binding sites, or with highly variable sites, can easily be missed by motif discovery algorithms. Furthermore, many proteins bind as hetero-dimers with multiple different pairs, and the motifs for different combinations may be somewhat different and perhaps not all have been sampled yet.

In a separate analysis, the DHSs within proximal-promoter regions (within 5 kb of the TSS) of all promoters were scanned using a set known PWMs for 475 TFs (Neph et al. 2012a). The promoter-proximal DHSs were used because they can be reliably assigned as regulatory sites for the adjacent gene, whereas DHSs that are far from any promoter have several possible genes that they may regulate. The matches that are found between TFs and their target genes provide a regulatory network, and inves-tigators were particularly interested in the connections between TFs, i.e., how TFs reg-ulate the expression of one another. The authors found an average of ~11,000 TF–TF edges in each of the 41 cell types. Figure 9-6 shows one example in which several TFs control the expression of IRF1, which in turn controls the expression of several other TFs. Several interesting findings emerge from the TF–TF networks that are observed.

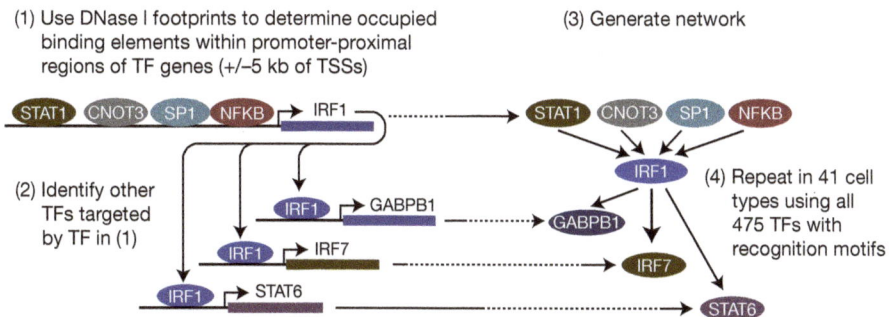

Figure 9-6. Regulatory networks between TFs. From DHS analysis and TF binding-site predictions using PWMs, target genes for many TFs are identified, allowing for the regulatory network between TFs to be inferred. This example shows that there are likely binding sites for several different TFs to con-trol expression of the IRF1 gene and that this in turn controls the expression of several other TFs. (Redrawn, with permission, from Neph et al. 2012a; © Elsevier.)

As expected, the networks vary among cell types. Networks found within individual cell types have many characteristics (such as FFLs being very common) that are observed in more well-studied organisms, such as bacteria and yeast. However, if all of the cell types are merged together into an overall TF–TF network, many of those characteristics are lost, such as the fact that FFLs are no longer overrepresented.

Alterations in regulatory networks that occur in cancer cell lines are often informative about the underlying dysregulation of gene expression associated with, and probably causative of, the cancer phenotype. Given the catalog of PWMs, the promoter-proximal regions can be scanned for matches and putative regulatory networks made from those. But the network characteristics that emerge from that analysis are quite different from those observed in real networks, and many of the predicted TF–TF connections have no experimental evidence. Most of the predictions based on PWM matches alone are likely false, as is commonly observed, because those sites are not accessible for protein binding. It is the combination of DHS mapping, which indicates which regions of the genome are accessible in any particular cell type, plus matches to PWMs, that together can lead to realistic regulatory networks. Currently, no algorithms can predict, from DNA sequence alone, the regions that will be accessible in any particular cell type; that information must be obtained experimentally. Perhaps someday, the rules will be well enough understood that regulatory networks can be predicted on sequence alone, but for now, the combination of DHS mapping and PWM predictions is likely to give the most comprehensive and accurate networks.

GRNs and Genetic Variation

Genetic variation—differences in DNA sequences—is important on two levels. One level is the differences among individuals of a species that give rise to the distinct individual features. For example, any two humans will be >99% identical across the entire genome, taking into account both single-nucleotide polymorphisms (SNPs) and other types of variation, such as deletions, duplications (collectively, copy-number variations [CNVs]), inversions, and translocations. It is the small number of differences that control many of the phenotypic differences among people (not including the differences due to environmental effects), such as hair, skin and eye color, height, and many others, including propensity for many diseases. There are many ongoing studies of the variation in human DNA sequences and how this variation is associated with different phenotypes and diseases, and many resources have been created from those studies and are publicly available (see Online Resources). The other level at which DNA variation is important is the differences among species. When two species initially separate from a single species, their genomes will be very similar, but over time, they will diverge, increasing the number of differences in their genomes along with increasing phenotypic differences. In 1965, Zuckerkandl and Pauling (1965) suggested that DNA differences among species could be used to measure the time since the species separated, essentially a molecular clock to be used for molecular phylogeny. At the time, only a few protein sequences were available for evidence, but the investigators assumed that the sequence from any inherited molecule, which they called a "semantide" (includes protein, DNA, and RNA sequences), would be useful. Molecular clocks have been found to have variable rates in different branches of the phylogenetic tree, but the overall concept of using molecular sequences, and especially DNA, for generating phylogenies, and even more importantly for studying the evolutionary events and processes that distinguish species, has been overwhelming, verified by the enormous increase in sequence information obtained since the publication of that paper.

Given that differences in DNA sequences can account for many of the differences among individuals within a species, as well as differences among species, an important determination is the mechanism by which those differences are manifested. There are two obvious and alternative types of changes, although they are not mutually exclusive: One is a change in gene content, either by the gain or loss of genes in one species compared to the other, or by modifications to genes such that they perform distinct, but perhaps overlapping, functions. The other is that gene content is unaffected, but the regulation of the genes changes, so that proteins are made in different amounts, at different times, in different cells, or under different conditions. Undoubtedly, both types of changes occur both within and between species, and it is important to determine which is more prevalent or more important for the particular observed differences.

In 1975, Mary-Claire King and Allan Wilson compared many proteins between humans and chimpanzees and concluded that the major differences between the species must be in how the genes are regulated rather than differences in gene functions (King and Wilson 1975). Efficient DNA sequencing methods were invented in 1977, and only since then could the issue really be addressed in detail. On the 30th anniversary of the King and Wilson paper, Sean Carroll published a brief review of the data collected in the intervening years that both supports the original idea and extends it in several directions (Carroll 2005). We previously described the use of phylogenetic footprinting (identifying the noncoding regions of the genome that are conserved through evolution) as an effective means of identifying the important regulatory sequences. But it is also clear that changes in gene regulation are important drivers of species-specific differences. Thus, identifying regulatory regions in one species that are missing or modified in another species is also an important aspect of understanding GRNs. We return to his later in the discussion of development regulation in *Drosophila* species, but the remainder of this section considers differences among individual humans that manifest as differences in gene regulation and contribute to differences in phenotypes and disease.

The concept of the "human genome" is really an abstraction that does not exist in reality. Each person has two different copies of the human genome, one obtained from each of their parents, and no two people have exactly the same genome sequence. Even identical twins, who were derived from a single zygote (fertilized egg), have been found to have a few differences in their DNA, presumably arising in the early stages of embryonic development. Even different cells within a single human will have variations in their DNA due to replication errors, most of which have no effect but some of which can lead to diseases such as cancer. There are many reasons for wanting to know the variation in DNA sequences among individual humans, and several large-scale projects have thus been undertaken (see Online Resources). This includes the early HapMap project to identify variations throughout the genome that could be used in studies of genotype–phenotype associations. More recent is the 1000 Genomes Project, now well on its way to sequencing 2000 individual genomes. Similar projects are being undertaken in several countries, and as genome sequencing becomes faster and less costly, there will be a large increase in the number of complete genomes sequences available for study. Cancer is a disease in which cells lose the normal control of their replication and continue dividing, and it is associated with, and probably caused by, changes to the DNA of individual cells. The Cancer Genome Anatomy Project (CGAP; see Online Resources) is sequencing individual cancers of several different types to identify the genes that are commonly associated with particular cancers and may provide diagnostic and/or therapeutic targets for the disease.

One important use of genetic variation data is in genome-wide association studies (GWAS), and there are many such studies ongoing (see Online Resources). In GWAS, one looks for genetic variations that are associated with specific phenotypes that may be normal variations among people, such as height, eye or hair color, etc., or with the occurrence of disease or risk of disease. For example, people may be classified into two groups—those with a particular disease and those without—and then investigators can determine the genomic variations in each group to try to detect those significantly correlated with one of the two classes. This can also be done with a quantitative trait, such as height, rather than a binary trait, such as having a disease or being healthy. In most studies, associated variants—most often SNPs—are identified, but it is unlikely that those are the causative variants responsible for the phenotypic differences. This is because SNPs (and other variations such as CNVs) tend to be correlated; that is, in linkage disequilibrium, with one another over variable distances along the DNA. Once an associated SNP is found, it is common to consider other correlated SNPs to attempt to identify the variation that causes the phenotype and that may lead to a better understanding of its molecular basis. Computational tools can facilitate the search for causative variants. For example, if the variation is expected to be due to a change in a protein, the correlated SNPs that occur within the coding sequence and are nonsynonymous (changes in the DNA that also change the amino acid sequence) would be considered, especially those that would be expected to affect protein structure and/or function. For example, changing an Ile to a Val (see Fig. 3-2B) is not likely to alter protein structure or function, whereas a change from an Ile to a Thr, going from a nonpolar hydrophobic amino acid to one that is a polar hydrophilic, is more likely to be disruptive. But a large majority of GWAS SNPs are noncoding and most likely to affect gene expression rather than gene function.

Gene expression itself is a phenotype that can be measured in cell cultures for collections of cells with known SNP variants, and in most cases, GWAS analysis identifies variations in the regulatory regions of genes as being responsible for the phenotype (Cheung et al. 2005). To go from those associated SNPs to the causative variants requires identifying not only SNPs that alter potential TF binding sites but also those that are likely to have an effect on binding. This is analogous to differentiating between synonymous and nonsynonymous protein changes. For example, if the consensus sequence for a TF is GTANCYG and the predicted binding site has the two variants GTAgCCG and GTAtCCG, that is unlikely to be a disruptive change because both the g and t variants match the consensus equally well at the N position. Likewise, two variants that both match the Y, C, and T are also unlikely to be disruptive, but a change to a G at that position would be predicted to alter the binding affinity and could be the causative variant. Instead of just using consensus sequences, PWMs can give more quantitative predictions about the effects of specific variants.

Several papers from the recent ENCODE collection as well as some prior papers have highlighted the ability of genome-wide binding-site analysis, both ChIP-seq and DHS, to identify likely causative noncoding variants from GWAS studies (Cooper and Shendure 2011; Pique-Regi et al. 2011; Boyle et al. 2012; Degner et al. 2012; Gaffney et al. 2012; Maurano et al. 2012a,b; Ni et al. 2012; Schaub et al. 2012; The ENCODE Project Consortium 2012; Vernot et al. 2012). Each of those papers includes examples of genome-wide binding-site analysis combined with information about specific TF motifs, generally PWMs, leading to likely alterations in TF binding that explain a significant GWAS result. In some cases, allele-specific expression, in which one of the two copies of the gene is expressed at a much higher level than the other copy, can also be explained by specific SNPs that change the binding site for a specific TF.

GRNs IN DEVELOPMENT

One of the most interesting and challenging problems in biology is the development of multicellular organisms. The initial single-celled zygote goes through many cell divisions to generate the complete organism that contains many different types of cells, each in the correct number and position in the organism and each expressing the subset of genes required for its specific cell type. A number of different organisms are commonly used to study development, each with its own advantages and disadvantages. To be most relevant to human biology, one would pick a mammalian system, and the mouse is the species of choice: It has a short generation time for a mammal and extensive genetic information, and a variety of well-characterized strains are available. Genetic manipulations are relatively easy, and there is a complete genome sequence and ENCODE project specifically on mouse (see Online Resources). Other vertebrate organisms that are commonly used are frogs, because of the large and easily obtained eggs facilitating both visualization of changes occurring during development and the ability to inject gene constructs and follow their expression through space and time, and zebrafish, because of their short generation time, good genetic tools, and ease of manipulation and visualization of cells, organs, and gene expression. The sea urchin has been a good model system for the study of GRNs for development in part because of the ease of getting large quantities of eggs and sperm, allowing for synchronous fertilization and developmental stages that are easy to visualize and in sufficient quantity for biochemical analysis. Recent advances in gene manipulation techniques and reporter assays and a complete genome sequence also facilitate its utility for studying development (see below).

From a genetic perspective, where the genes required for normal development and the type of defects that occur when those genes are nonfunctional need to be identified, flies and worms are the most commonly used organisms. They have short generation times, are easily crossed to perform genetics experiments, and have relatively small genomes (~100 Mb) that have been sequenced, including several closely related species. Besides many existing mutations that affect development, easy methods are available to do genetic manipulations and determine the effects of specific mutations. The following sections briefly describe some of what is known about GRNs in the early stages of sea urchin and fly development.

Sea Urchin

Sea urchins have been studied extensively for GRNs that control development, with much of the molecular information obtained from the species *Strongylocentrotus purpuratus* (McClay 2011). Eggs and sperm can be readily collected and added together to get synchronous fertilization and development in large batches suitable for biochemical analysis. Injections can be performed to follow reporter genes as well as to knock down expression of specific genes. Differentiation begins rapidly with visibly distinct cell types by the fourth cell division (16-cell stage), which occurs ~4 h after fertilization. Within 24 h, the embryo is in the blastula stage and after 2 d, the gastrula has formed, with the complete larva obtained by the third day. The larva contains many differentiated cell types that can be separated relatively easily for independent biochemical analysis. Regulatory regions for a large number of genes have been identified and, in many cases, so have the TFs that bind to them. A few studies from the Davidson laboratory related to identifying critical connections in the GRN for early development are listed here (Yuh et al. 1998, 2001; Davidson et al. 2002; Oliveri et al. 2008; Nam et al. 2010). A website and online database, spbase (see Online Resources), are continually updated with information about the developmental regulatory network, with tools available for visualizing and performing simulations on the network or its subnetworks. Figure 9-7 shows an example of a small section of the overall network that gives rise to the skeletogenic mesoderm lineage. The entire network, including various time points during development, is available from the website. Figure 9-7 shows the network at a high level, linking TFs to the genes that they regulate, but extensive information is also known at the DNA sequence level, indicating where specific TFs bind. Figure 9-8A shows the upstream regulatory region for the *endo16* gene, divided into segments A through G, which are used at different times and in different tissues to control the expression of *endo16* (Yuh et al. 1998, 2001). Figure 9-8B is an enlargement of Fig. 9-8A, showing the specific binding sites for individual TFs that are involved in direct regulation of *endo16*, and Figure 9-8C shows some of the combinatorial logic from the set of upstream segments that control expression in different tissues.

Drosophila Embryonic Patterning

The best-studied developmental regulatory network is for the *Drosophila melanogaster* early embryo when it establishes gene expression patterns that are further elaborated to give rise to the complete animal. *Drosophila* has been an important model organism for many different biological processes for more than a century. It was the first species for which a genetic map was constructed, and it was in *Drosophila* that it was shown that genes are associated with chromosomes, both achieved in the early 20th century by T.H. Morgan's "Fly Room" laboratory (and for which Morgan won the 1933 Nobel Prize in Physiology or Medicine). But work first published by Christiane Nusslein-Volhard and Eric Wieschaus (1980), for which they were awarded—along with Edward Lewis—the 1995 Nobel Prize in Medicine or Physiology, established the paradigm for determining the GRN controlling vital developmental processes. By mutagenizing flies and screening for death of early embryos, these researchers identified 15 genes that were essential for normal embryonic development. By also considering the stage at which the embryo died, they identified three stages of development and could classify each gene as affecting one of those stages. In the intervening years, much more has been learned about additional genes involved and the regulatory network that controls gene expression in the early development of the fly embryo. A variety of technological advances facilitate the manipulation of genes, the localization of proteins and mRNAs within the embryo, and the identification and assessment of regulatory regions. The *D. melanogaster* genome has been sequenced as have the genomes from several related species, and the modENCODE project has contributed additional information about the TFs and their binding sites across the genome. The following paragraphs provide a very brief description of the initial steps of embryonic development and then describe the specific regulation of one gene.

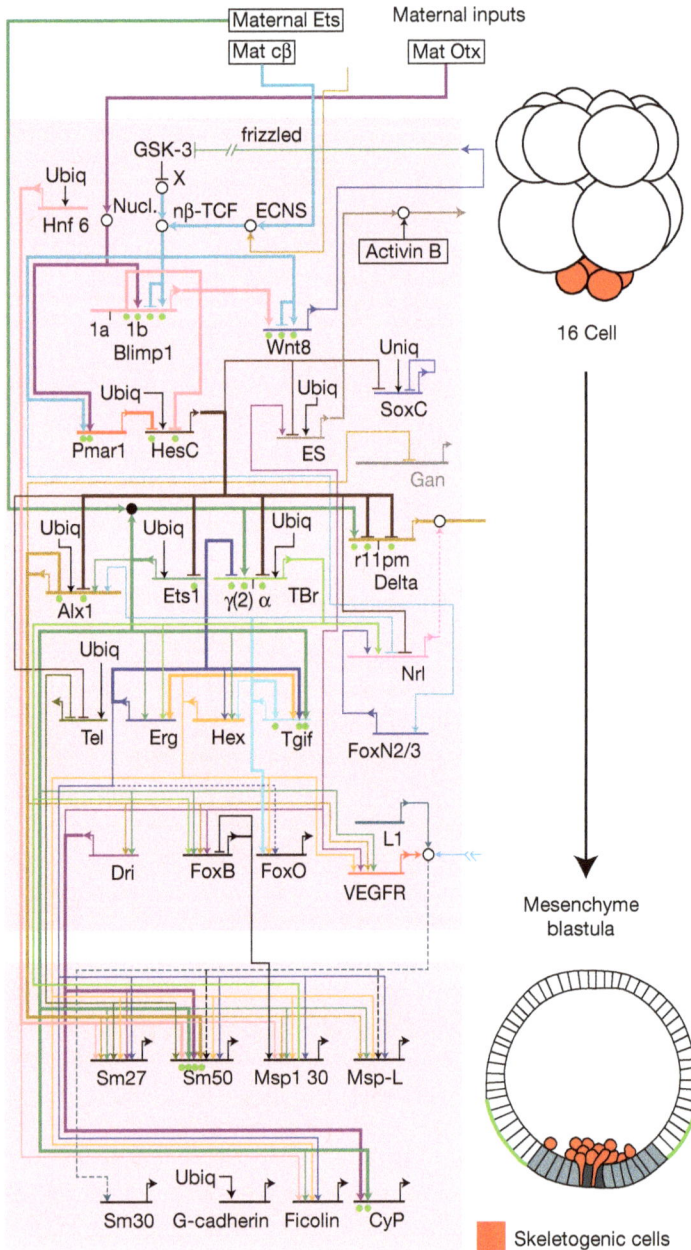

Figure 9-7. A regulatory network from sea urchin. The network is shown for many genes involved in skeletogenic micromere specification. (*Top line*) Maternal TFs that initiate the network beginning at the 16-cell stage (red cells in the "embryo" image, *top right*). Activation of the remainder of the genes shown occurs over time, requiring direct input from previously expressed genes as indicated by the TF edges. The genes shown at the bottom of the network are active at the time of the mesenchyme blastula stage, at which point the skeletogenic cells are specified (red cells, *lower right*). (Reproduced, with permission, from McClay 2011; © Company of Biologists.)

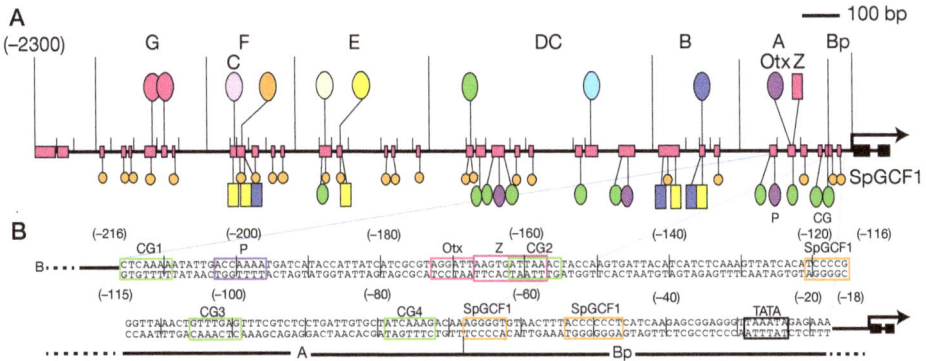

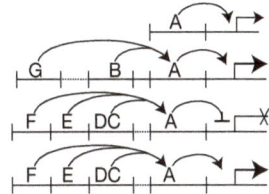

Figure 9-8. Regulation of *endo16* in different cells and at different times. (A) Upstream region of *endo16* gene containing the regulatory regions controlling its expression. The entire region, of ~2.3 kb, is divided into segments A–G that specify different functions. The binding sites for distinct TFs are shown as different-colored ovals and rectangles. (B) Enlargement of segment A and the promoter indicating the binding specific sites for several of the regulatory TFs. (C) Logic of *endo16* regulation, indicating how the occupancy of different segments by their TFs controls expression in different cells and at different times. (Reproduced, with permission, from Yuh et al. 1998; © AAAS.)

The *Drosophila* fertilized egg goes through a series of 10 rapid nuclear divisions generating ~1000 nuclei in ~80 min. Most of the nuclei then migrate to the periphery of the embryo and nuclear transcription begins. At this point, there are no cells, just the collection of nuclei, and their initial transcription is governed by TFs that were donated to the egg by the mother. The maternal TFs that direct early transcription, such as bicoid, hunchback, and caudal, are not evenly distributed throughout the embryo but occur in gradients; bicoid and hunchback are highly expressed in the anterior part of the embryo, but with somewhat different patterns, and caudal is highly expressed in the posterior end. The first zygotic genes expressed are gap genes, which are TFs whose expression is controlled by maternal genes as well as by one another. Mutations in gap genes lead to a loss of embryonic segments, and later experiments showed that each are expressed in distinct striped patterns along the anterior–posterior (A–P) axis of the embryo. Next is the expression of pair-rule genes, which are controlled by a combination of maternal TFs, gap genes, and one another; in fact, many of them have autoregulatory connections. They are also TFs and are required to define the segment boundaries with mutations in those genes leading to the reduction of specific segments. For example, mutations in the *even-skipped* (*eve*) gene eliminate every other (the even-numbered) segments of the embryo. The remaining class of genes is segment polarity genes, which are TFs that define the A–P polarity of each segment and are each controlled by specific combinations of the expressed TFs. The expression of each class of genes occurs between nuclear division 10, when zygotic transcription begins, and nuclear division 14, between ~80 min and 130 min after fertilization. By nuclear division 14, cells are formed around each nucleus and their divisions become asynchronous. Before cellularization, when all nuclei share the same cytoplasm (which is the entire embryo), TFs that are made from one nucleus can affect expression in other nuclei, limited only by their diffusion rates. Because initial maternal TFs are in gradients and the genes that they regulate (and all subsequent genes) are differentially sensitive to different TF concentrations, each new round of TF expression is also spatially patterned, giving rise to the specific segmentation pattern, with different segments having unique identities and leading to different body parts of the adult fly. In the mid 1980s, techniques were developed that used fluorescent antibodies to determine the spatial localization of specific TFs within the embryo. The pair-rule gene *fushi tarazu* (*ftz*) was found to be localized in seven stripes along the A–P axis, and the other genes required for that pattern were identified by observing the localization pattern in mutants of other TFs (Carroll and Scott 1985, 1986). *eve* was also discovered to be expressed in seven stripes that were complementary to the pattern of *ftz* and controlled by a common set of TFs (Frasch and Levine 1987; Frasch et al. 1987).

In recent years, automated imaging and image alignment techniques have greatly increased the ability to collect high-resolution data, both in time and space, for both protein and mRNA localization patterns (Myasnikova et al. 2001; Fowlkes et al. 2008), and data for many different proteins, from multiple species and in various genetic backgrounds, are now available in online databases such as FlyNet and Fly-Ex (see Online Resources). A variety of experimental approaches have also been applied that identify which TFs control the expression of each gene and in which the regulatory regions and binding sites are located for each gene. This is a perfect system for developing and testing computational models of GRNs to see whether observed spatial and temporal dynamics can be well modeled and the effects of mutations to either TFs or specific binding sites accurately predicted. This has been an active area of research during the last 20 years, and a variety of computational methods has been developed and continues to be improved (Reinitz et al. 1995, 1998; Jaeger et al. 2004; Janssens et al. 2006; Segal et al. 2008; MacArthur et al. 2009; Kazemian et al. 2010; Schroeder et al. 2011; Wunderlich et al. 2012).

The entire regulatory network is now well characterized, with the regulatory regions for each TF mapped and the binding sites for the regulatory TFs known. There are also extensive comparisons with other species and generally, when those regulatory regions are moved into *D. melanogaster*, an identical or very similar pattern of expression is observed. This is true even if the individual TF binding sites have been rearranged extensively so that all regulatory regions are not significantly alignable using standard dynamic programming methods (see Box 8-1). As discussed above, one of the best-characterized genes is *eve*, a pair-rule gene that is expressed in seven stripes in the early embryo and then is also expressed in additional locations later in development (Frasch et al. 1987). The set of TFs required for normal *eve* expression has been determined, and a regulatory network showing which affect transcription positively or negatively can be drawn (Fig. 9-9A). This map appears very complicated, with a large number of TFs contributing to the complex expression pattern of *eve*. But there are many separable regulatory regions and when they are considered individually the picture becomes much simpler (Fig. 9-9B). For example, *eve* stripe 2 is controlled by a region of ∼500 bp that contains binding sites for just four maternal and gap genes: *bicoid* (five binding sites), *hunchback* (one binding site), *giant* (three binding sites), and *Kruppel* (three binding sites) (Small et al. 1991, 1992; Stanojevic et al. 1991; Arnosti et al. 1996). If that particular regulatory region (enhancer) is used to drive expression of a reporter gene, it is expressed only in the *eve* stripe 2 pattern. Furthermore, if that particular enhancer is deleted or mutated, only the *eve* stripe 2 is missing or modified; the other stripes are unchanged. Enhancers for *eve* expression have been located at various locations relative to the gene; for example, the *eve* stripe 2 enhancer is just upstream of the TSS, whereas others occur downstream from the gene. Individual enhancers are controlled by only one to five different TFs. The specific expression pattern is determined by a combination of the binding sites within each enhancer and the concentration of the TFs that bind that enhancer at the particular location in the embryo.

As described above, the function of enhancers is conserved across *Drosophila* species. If one looks at even more distantly related species, such as sepsid flies, which diverged from *Drosophila* ∼100 million years ago (longer than the divergence between human and mouse), functional conservation can still be observed even though the sequences are highly diverged (Hare et al. 2008a,b). Early TFs whose expression patterns have been measured are essentially the same among species, and the corresponding enhancers can be identified by their locations relative to the genes. For example, the region just upstream of the sepsid *eve* gene, when used to drive a reporter gene in *Drosophila*, gives an expression pattern nearly identical to the *eve* stripe 2 enhancer. When the sequences are compared, there are a few 20- to 30-bp blocks of high conservation, but overall, the sequences are quite dissimilar. The conserved blocks correspond to the TF binding sites, and generally to pairs of binding sites, in the *Drosophila* enhancer. Figure 9-10 shows the corresponding *eve* stripe 2 enhancers from four *Drosophila* species and four sepsid species, with experimentally determined and computationally predicted (based on PWM scores) binding sites for each of the four TFs that control expression from that enhancer. It is evident that all of the enhancers are enriched for those binding sites, and several adjacent pairs are conserved, but the overall arrangements are quite different. The sepsid

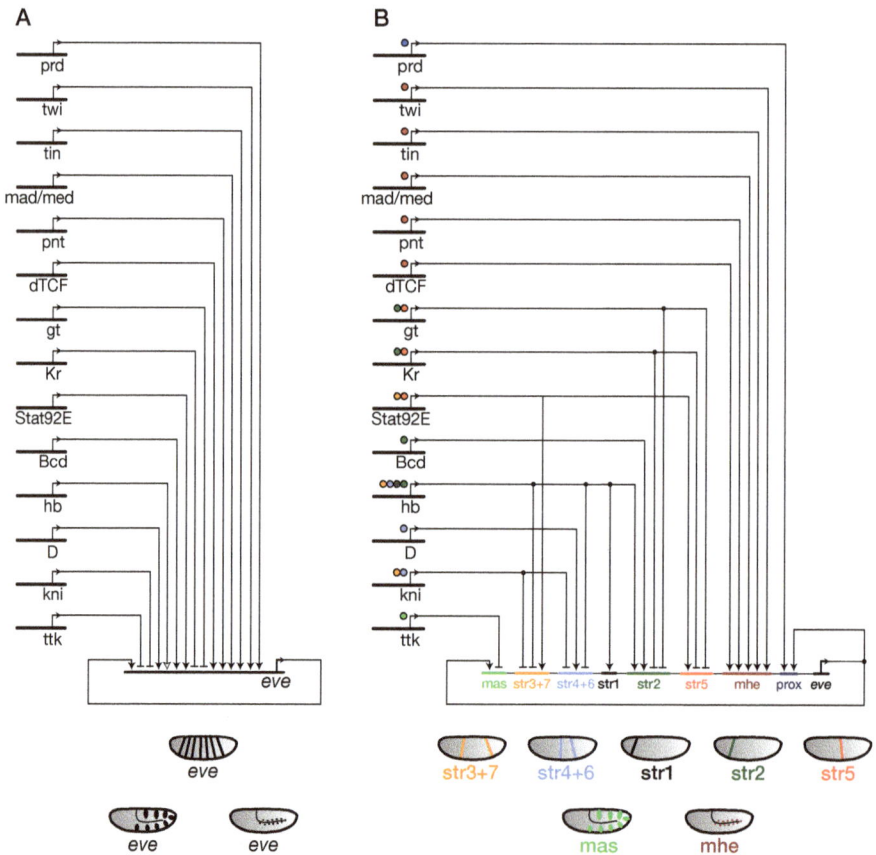

Figure 9-9. Regulatory input to the *eve* gene of *Drosophila*. (*A, left column*) Set of TFs that contribute to *eve* gene expression. Arrows or bars along the *eve* gene indicate whether they activate or repress transcription, respectively. (*Hunchback* represses in some locations and activates in others.) Note that *eve* is also autoregulatory. (*A, bottom*) Pattern of *eve* expression in the embryo. At early times, it is expressed in seven distinct stripes, whereas at later times it is expressed in heart cells and some neurons. (*B*) Different portions of the expression pattern controlled by TFs binding to independent enhancer regions. Shown are TFs that bind to each of the independent enhancers. (*B, below*) Pattern of gene expression derived from that enhancer. Each enhancer has one to five TFs as input and drives expression of either one or two stripes and either the heart or neuronal cells. (Prox indicates proximal promoter elements at the start of transcription.) (Reproduced, with permission, from Wilczynski and Furlong 2010; © Elsevier.)

enhancer regions are longer overall and the sites are more spread out. The functional conservation indicates that the overall density of sites, and probably their local clustering, is the essential feature and in other respects the sequences can diverge without detrimental effect.

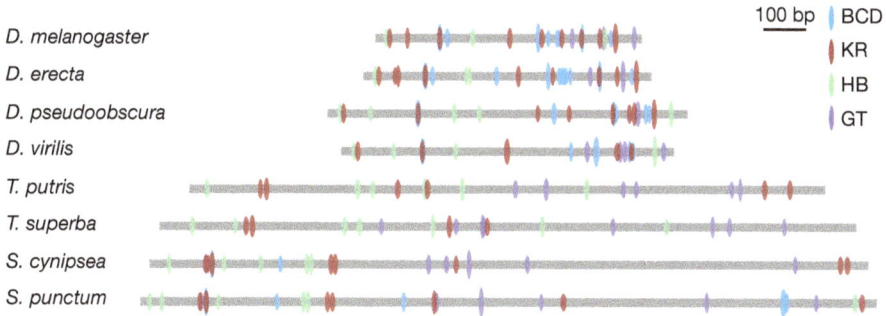

Figure 9-10. Evolution of the *eve* stripe 2 enhancers, shown for four *Drosophila* species and four sepsid fly species. (Different-colored ovals) Binding sites for the four TFs bicoid (BCD), hunchback (HB), giant (GT), and Kruppel (KR). Binding sites in *Drosophila*, and some from other species, are from experimental data; the rest are predicted based on the PWM for each protein. In all cases, the height of the oval corresponds to the PWM score, which is likely to be related to its binding affinity. (Reproduced from Hare et al. 2008a.)

CONCLUSIONS

The interactions of TFs with DNA are now recognized to be important in most fields of biology. The control of gene expression is fundamental to the characteristics and behavior of all cells, and misregulation often causes disease or other dysfunctional phenotypes. Most of this book has focused on the intrinsic properties of the interactions from various viewpoints. By observing, measuring, and modeling the direct binding interaction, without the confounding effects that can occur in vivo, we can gain insights into the primary molecular principles of specificity. Such an understanding is worthwhile on its own, but it also facilitates accurate predictions of the consequences of variation: What happens when the protein sequence or DNA-binding site is altered and can we use that knowledge in the design of systems with desirable characteristics? When applied to in vivo biology, as in the many examples briefly described in this chapter, the knowledge of intrinsic specificity is both limited and enabling. It is limited because, by itself, this knowledge leads to many false-positive predictions. If we try to predict the regulatory sites in a genome based on PWM scores, the vast majority of the sites will not be functional because we are missing critical knowledge about their accessibility. If we have both accessibility information and specific models, the combined knowledge is synergistic, being more valuable than the sum of their independent contributions.

The same is true of genetic variation analysis, both within species, where it is useful for modeling phenotypic variation and disease association, and between species, where it provides insights into the processes that are both common and distinctive. Furthermore, having reliable models of intrinsic specificity makes false-negative predictions very valuable. If it can be predicted confidently that a TF would not bind to a sequence where the experiments clearly show that it does, that discrepancy points to a critical piece of missing information, most likely an indirect binding event due to some intermediate factor, and it directs the search for the missing connection. Without reliable models of specificity, such searches could easily be mistaken. Information about intrinsic specificity, both from experimental determinations and from computational models, is already making important contributions to our understanding biology and many aspects of medicine. We anticipate that the contributions will increase as the gaps in our knowledge diminish.

REFERENCES

Aerts S. 2012. Computational strategies for the genome-wide identification of cis-regulatory elements and transcriptional targets. *Curr Top Dev Biol* **98:** 121–145.

Alon U. 2007. Network motifs: Theory and experimental approaches. *Nat Rev Genet* **8:** 450–461.

Arnosti DN, Barolo S, Levine M, Small S. 1996. The eve stripe 2 enhancer employs multiple modes of transcriptional synergy. *Development* **122:** 205–214.

Babu MM. 2008. Computational approaches to study transcriptional regulation. *Biochem Soc Trans* **36:** 758–765.

Boyle AP, Hong EL, Hariharan M, Cheng Y, Schaub MA, Kasowski M, Karczewski KJ, Park J, Hitz BC, Weng S, et al. 2012. Annotation of functional variation in personal genomes using RegulomeDB. *Genome Res* **22:** 1790–1797.

Carroll SB. 2005. Evolution at two levels: On genes and form. *PLoS Biol* **3:** e245.

Carroll SB, Scott MP. 1985. Localization of the fushi tarazu protein during *Drosophila* embryogenesis. *Cell* **43:** 47–57.

Carroll SB, Scott MP. 1986. Zygotically active genes that affect the spatial expression of the *fushi tarazu* segmentation gene during early *Drosophila* embryogenesis. *Cell* **45:** 113–126.

Cheung VG, Spielman RS, Ewens KG, Weber TM, Morley M, Burdick JT. 2005. Mapping determinants of human gene expression by regional and genome-wide association. *Nature* **437:** 1365–1369.

Cooper GM, Shendure J. 2011. Needles in stacks of needles: Finding disease-causal variants in a wealth of genomic data. *Nat Rev Genet* **12:** 628–640.

Davidson EH, Rast JP, Oliveri P, Ransick A, Calestani C, Yuh CH, Minokawa T, Amore G, Hinman V, Arenas-Mena C, et al. 2002. A genomic regulatory network for development. *Science* **295:** 1669–1678.

Degner JF, Pai AA, Pique-Regi R, Veyrieras JB, Gaffney DJ, Pickrell JK, De Leon S, Michelini K, Lewellen N, Crawford GE, et al. 2012. DNase I sensitivity QTLs are a major determinant of human expression variation. *Nature* **482:** 390–394.

Dehal PS, Joachimiak MP, Price MN, Bates JT, Baumohl JK, Chivian D, Friedland GD, Huang KH, Keller K, Novichkov PS, et al. 2010. MicrobesOnline: An integrated portal for comparative and functional genomics. *Nucleic Acids Res* **38:** D396–D400.

de Hoon MJ, Eichenberger P, Vitkup D. 2010. Hierarchical evolution of the bacterial sporulation network. *Curr Biol* **20:** R735–R745.

Drubin DA, Way JC, Silver PA. 2007. Designing biological systems. *Genes Dev* **21:** 242–254.

Ecker JR, Bickmore WA, Barroso I, Pritichard JK, Gilad Y, Segal E. 2012. Genomics: ENCODE explained. *Nature* **489:** 52–55.

Elowitz MB, Leibler S. 2000. A synthetic oscillatory network of transcriptional regulators. *Nature* **403:** 335–338.

Eser U, Falleur-Fettig M, Johnson A, Skotheim JM. 2011. Commitment to a cellular transition precedes genome-wide transcriptional change. *Mol Cell* **43:** 515–527.

Fowlkes CC, Hendriks CL, Keranen SV, Weber GH, Rübel O, Huang MY, Chatoor S, DePace AH, Simirenko L, Henriquez C, et al. 2008. A quantitative spatiotemporal atlas of gene expression in the *Drosophila* blastoderm. *Cell* **133:** 364–374.

Frasch M, Levine M. 1987. Complementary patterns of even-skipped and fushi tarazu expression involve their differential regulation by a common set of segmentation genes in *Drosophila*. *Genes Dev* **1:** 981–995.

Frasch M, Hoey T, Rushlow C, Doyle H, Levine M. 1987. Characterization and localization of the even-skipped protein of *Drosophila*. *EMBO J* **6:** 749–759.

Gaffney DJ, Veyrieras JB, Degner JF, Pique-Regi R, Pai AA, Crawford GE, Stephens M, Gilad Y, Pritchard JK. 2012. Dissecting the regulatory architecture of gene expression QTLs. *Genome Biol* **13:** R7.

Gardner TS, Cantor CR, Collins JJ. 2000. Construction of a genetic toggle switch in *Escherichia coli*. *Nature* **403:** 339–342.

Gerstein MB, Kundaje A, Hariharan M, Landt SG, Yan KK, Cheng C, Mu XJ, Khurana E, Rozowsky J, Alexander R, et al. 2012. Architecture of the human regulatory network derived from ENCODE data. *Nature* **489:** 91–100.

Ghosh S, Matsuoka Y, Asai Y, Hsin KY, Kitano H. 2011. Software for systems biology: From tools to integrated platforms. *Nat Rev Genet* **12:** 821–832.

Giresi PG, Kim J, McDaniell RM, Iyer VR, Lieb JD. 2007. FAIRE (formaldehyde-assisted isolation of regulatory elements) isolates active regulatory elements from human chromatin. *Genome Res* **17:** 877–885.

Gruber TM, Gross CA. 2003. Multiple sigma subunits and the partitioning of bacterial transcription space. *Annu Rev Microbiol* **57:** 441–466.

Guido NJ, Wang X, Adalsteinsson D, McMillen D, Hasty J, Cantor CR, Elston TC, Collins JJ. 2006. A bottom-up approach to gene regulation. *Nature* **439:** 856–860.

Harbison CT, Gordon DB, Lee TI, Rinaldi NJ, Macisaac KD, Danford TW, Hannett NM, Tagne JB, Reynolds DB, Yoo J, et al. 2004. Transcriptional regulatory code of a eukaryotic genome. *Nature* **431:** 99–104.

Hare EE, Peterson BK, Eisen MB. 2008a. A careful look at binding site reorganization in the even-skipped enhancers of *Drosophila* and sepsids. *PLoS Genet* **4:** e1000268.

Hare EE, Peterson BK, Iyer VN, Meier R, Eisen MB. 2008b. Sepsid even-skipped enhancers are functionally conserved in *Drosophila* despite lack of sequence conservation. *PLoS Genet* **4:** e1000106.

Hasty J, McMillen D, Isaacs F, Collins JJ. 2001. Computational studies of gene regulatory networks: In numero molecular biology. *Nat Rev Genet* **2:** 268–279.

Hecker M, Lambeck S, Toepfer S, van Someren E, Guthke R. 2009. Gene regulatory network inference: Data integration in dynamic models—A review. *Biosystems* **96:** 86–103.

Herrgard MJ, Covert MW, Palsson BO. 2004. Reconstruction of microbial transcriptional regulatory networks. *Curr Opin Biotechnol* **15:** 70–77.

Hesselberth JR, Chen X, Zhang Z, Sabo PJ, Sandstrom R, Reynolds AP, Thurman RE, Neph S, Kuehn MS, Noble WS, et al. 2009. Global mapping of protein-DNA interactions in vivo by digital genomic footprinting. *Nat Methods* **6:** 283–289.

Jaeger J, Surkova S, Blagov M, Janssens H, Kosman D, Kozlov KN, Manu, Myasnikova E, Vanario-Alonso CE, Samsonova M, et al. 2004. Dynamic control of positional information in the early *Drosophila* embryo. *Nature* **430:** 368–371.

Janga SC, Collado-Vides J. 2007. Structure and evolution of gene regulatory networks in microbial genomes. *Res Microbiol* **158:** 787–794.

Janky R, Helden J, Babu MM. 2009. Investigating transcriptional regulation: From analysis of complex networks to discovery of cis-regulatory elements. *Methods* **48:** 277–286.

Janssens H, Hou S, Jaeger J, Kim AR, Myasnikova E, Sharp D, Reinitz J. 2006. Quantitative and predictive model of transcriptional control of the *Drosophila melanogaster* even skipped gene. *Nat Genet* **38:** 1159–1165.

Johnson DS, Mortazavi A, Myers RM, Wold B. 2007. Genome-wide mapping of in vivo protein-DNA interactions. *Science* **316:** 1497–1502.

Kaern M, Blake WJ, Collins JJ. 2003. The engineering of gene regulatory networks. *Annu Rev Biomed Eng* **5:** 179–206.

Kazakov AE, Cipriano MJ, Novichkov PS, Minovitsky S, Vinogradov DV, Arkin A, Mironov AA, Gelfand MS, Dubchak I. 2007. RegTransBase—A database of regulatory sequences and interactions in a wide range of prokaryotic genomes. *Nucleic Acids Res* **35:** D407–D412.

Kazemian M, Blatti C, Richards A, McCutchan M, Wakabayashi-Ito N, Hammonds AS, Celniker SE, Kumar S, Wolfe SA, Brodsky MH, Sinha S. 2010. Quantitative analysis of the *Drosophila* segmentation regulatory network using pattern generating potentials. *PLoS Biol* **8.** pii.

Keseler IM, Collado-Vides J, Santos-Zavaleta A, Peralta-Gil M, Gama-Castro S, Muñiz-Rascado L, Bonavides-Martinez C, Paley S, Krummenacker M, Altman T, et al. 2011. EcoCyc: A comprehensive database of *Escherichia coli* biology. *Nucleic Acids Res* **39:** D583–D590.

Khalil AS, Collins JJ. 2010. Synthetic biology: Applications come of age. *Nat Rev Genet* **11:** 367–379.

Khalil AS, Lu TK, Bashor CJ, Ramirez CL, Pyenson NC, Joung JK, Collins JJ. 2012. A synthetic biology framework for programming eukaryotic transcription functions. *Cell* **150:** 647–658.

Kim TM, Park PJ. 2011. Advances in analysis of transcriptional regulatory networks. *Wiley Interdiscip Rev Syst Biol Med* **3:** 21–35.

King MC, Wilson AC. 1975. Evolution at two levels in humans and chimpanzees. *Science* **188:** 107–116.

Lawrence CE, Reilly AA. 1990. An expectation maximization (EM) algorithm for the identification and characterization of common sites in unaligned biopolymer sequences. *Proteins* **7:** 41–51.

Lee TI, Rinaldi NJ, Robert F, Odom DT, Bar-Joseph Z, Gerber GK, Hannett NM, Harbison CT, Thompson CM, Simon I, et al. 2002. Transcriptional regulatory networks in *Saccharomyces cerevisiae*. *Science* **298:** 799–804.

MacArthur S, Li XY, Li J, Brown JB, Chu HC, Zeng L, Grondona BP, Hechmer A, Simirenko L, Keränen SV, et al. 2009. Developmental roles of 21 *Drosophila* transcription factors are determined by quantitative differences in binding to an overlapping set of thousands of genomic regions. *Genome Biol* **10:** R80.

Maurano MT, Humbert R, Rynes E, Thurman RE, Haugen E, Wang H, Reynolds AP, Sandstrom R, Qu H, Brody J, et al. 2012a. Systematic localization of common disease-associated variation in regulatory DNA. *Science* **337:** 1190–1195.

Maurano MT, Wang H, Kutyavin T, Stamatoyannopoulos JA. 2012b. Widespread site-dependent buffering of human regulatory polymorphism. *PLoS Genet* **8:** e1002599.

McClay DR. 2011. Evolutionary crossroads in developmental biology: Sea urchins. *Development* **138:** 2639–2648.

Melnikov A, Murugan A, Zhang X, Tesileanu T, Wang L, Rogov P, Feizi S, Gnirke A, Callan CG Jr, Kinney JB, et al. 2012. Systematic dissection and optimization of inducible enhancers in human cells using a massively parallel reporter assay. *Nat Biotechnol* **30:** 271–277.

Mukherji S, van Oudenaarden A. 2009. Synthetic biology: Understanding biological design from synthetic circuits. *Nat Rev Genet* **10:** 859–871.

Myasnikova E, Samsonova A, Kozlov K, Samsonova M, Reinitz J. 2001. Registration of the expression patterns of *Drosophila* segmentation genes by two independent methods. *Bioinformatics* **17:** 3–12.

Nam J, Dong P, Tarpine R, Istrail S, Davidson EH. 2010. Functional cis-regulatory genomics for systems biology. *Proc Natl Acad Sci* **107:** 3930–3935.

Neph S, Stergachis AB, Reynolds A, Sandstrom R, Borenstein E, Stamatoyannopoulos JA. 2012a. Circuitry and dynamics of human transcription factor regulatory networks. *Cell* **150:** 1274–1286.

Neph S, Vierstra J, Stergachis AB, Reynolds AP, Haugen E, Vernot B, Thurman RE, John S, Sandstrom R, Johnson AK, et al. 2012b. An expansive human regulatory lexicon encoded in transcription factor footprints. *Nature* **489:** 83–90.

Ni Y, Hall AW, Battenhouse A, Iyer VR. 2012. Simultaneous SNP identification and assessment of allele-specific bias from ChIP-seq data. *BMC Genet* **13:** 79.

Nusslein-Volhard C, Wieschaus E. 1980. Mutations affecting segment number and polarity in *Drosophila*. *Nature* **287:** 795–801.

Oliveri P, Tu Q, Davidson EH. 2008. Global regulatory logic for specification of an embryonic cell lineage. *Proc Natl Acad Sci* **105:** 5955–5962.

Patwardhan RP, Hiatt JB, Witten DM, Kim MJ, Smith RP, May D, Lee C, Andrie JM, Lee SI, Cooper GM, et al. 2012. Massively parallel functional dissection of mammalian enhancers in vivo. *Nat Biotechnol* **30:** 265–270.

Perez JC, Groisman EA. 2009. Evolution of transcriptional regulatory circuits in bacteria. *Cell* **138:** 233–244.

Pique-Regi R, Degner JF, Pai AA, Gaffney DJ, Gilad Y, Pritchard K. 2011. Accurate inference of transcription factor binding from DNA sequence and chromatin accessibility data. *Genome Res* **21:** 447–455.

Reinitz J, Mjolsness E, Sharp DH. 1995. Model for cooperative control of positional information in *Drosophila* by bicoid and maternal hunchback. *J Exp Zool* **271:** 47–56.

Reinitz J, Kosman D, Vanario-Alonso CE, Sharp DH. 1998. Stripe forming architecture of the gap gene system. *Dev Genet* **23:** 11–27.

Ro DK, Paradise EM, Ouellet M, Fisher KJ, Newman KL, Ndungu JM, Ho KA, Eachus RA, Ham TS, Kirby J, et al. 2006. Production of the antimalarial drug precursor artemisinic acid in engineered yeast. *Nature* **440:** 940–943.

Roth FP, Hughes JD, Estep PW, Church GM. 1998. Finding DNA regulatory motifs within unaligned noncoding sequences clustered by whole-genome mRNA quantitation. *Nat Biotechnol* **16:** 939–945.

Sahota G, Stormo GD. 2010. Novel sequence-based method for identifying transcription factor binding sites in prokaryotic genomes. *Bioinformatics* **26:** 2672–2677.

Salgado H, Martinez-Flores I, Lopez-Fuentes A, Garcia-Sotelo JS, Porrón-Sotelo L, Solano H, Muñiz-Rascado L, Collado-Vides J. 2012. Extracting regulatory networks of *Escherichia coli* from RegulonDB. *Methods Mol Biol* **804:** 179–195.

Schaub MA, Boyle AP, Kundaje A, Batzoglou S, Snyder M. 2012. Linking disease associations with regulatory information in the human genome. *Genome Res* **22:** 1748–1759.

Schena M, Shalon D, Davis RW, Brown PO. 1995. Quantitative monitoring of gene expression patterns with a complementary DNA microarray. *Science* **270:** 467–470.

Schroeder MD, Greer C, Gaul U. 2011. How to make stripes: Deciphering the transition from non-periodic to periodic patterns in *Drosophila* segmentation. *Development* **138:** 3067–3078.

Segal E, Raveh-Sadka T, Schroeder M, Unnerstall U, Gaul U. 2008. Predicting expression patterns from regulatory sequence in *Drosophila* segmentation. *Nature* **451:** 535–540.

Sierro N, Makita Y, de Hoon M, Nakai K. 2008. DBTBS: A database of transcriptional regulation in *Bacillus subtilis* containing upstream intergenic conservation information. *Nucleic Acids Res* **36:** D93–D96.

Small S, Kraut R, Hoey T, Warrior R, Levine M. 1991. Transcriptional regulation of a pair-rule stripe in *Drosophila*. *Genes Dev* **5:** 827–839.

Small S, Blair A, Levine M. 1992. Regulation of even-skipped stripe 2 in the *Drosophila* embryo. *EMBO J* **11:** 4047–4057.

Stamatoyannopoulos JA. 2012. What does our genome encode? *Genome Res* **22:** 1602–1611.

Stanojevic D, Small S, Levine M. 1991. Regulation of a segmentation stripe by overlapping activators and repressors in the *Drosophila* embryo. *Science* **254:** 1385–1387.

Stormo GD, Hartzell GW 3rd. 1989. Identifying protein-binding sites from unaligned DNA fragments. *Proc Natl Acad Sci* **86:** 1183–1187.

The ENCODE Project Consortium. 2012. An integrated encyclopedia of DNA elements in the human genome. *Nature* **489:** 57–74.

Thurman RE, Rynes E, Humbert R, Vierstra J, Maurano MT, Haugen E, Sheffield NC, Stergachis AB, Wang H, Vernot B, et al. 2012. The accessible chromatin landscape of the human genome. *Nature* **489:** 75–82.

van Steensel B, Henikoff S. 2000. Identification of in vivo DNA targets of chromatin proteins using tethered dam methyltransferase. *Nat Biotechnol* **18:** 424–428.

Vernot B, Stergachis AB, Maurano MT, Vierstra J, Neph S, Thurman RE, Stamatoyannopoulos JA, Akey JM, et al. 2012. Personal and population genomics of human regulatory variation. *Genome Res* **22:** 1689–1697.

Walhout AJ. 2006. Unraveling transcription regulatory networks by protein-DNA and protein-protein interaction mapping. *Genome Res* **16:** 1445–1454.

Wang Z, Gerstein M, Snyder M. 2009. RNA-Seq: A revolutionary tool for transcriptomics. *Nat Rev Genet* **10:** 57–63.

Wang H, Mayhew D, Chen X, Johnston M, Mitra RD. 2011. Calling cards enable multiplexed identification of the genomic targets of DNA-binding proteins. *Genome Res* **21:** 748–755.

Wang H, Mayhew D, Chen X, Johnston M, Mitra RD. 2012a. "Calling cards" for DNA-binding proteins in mammalian cells. *Genetics* **190:** 941–949.

Wang J, Zhuang J, Iyer S, Lin X, Whitfield TW, Greven MC, Pierce BG, Dong X, Kundaje A, Cheng Y, et al. 2012b. Sequence features and chromatin structure around the genomic regions bound by 119 human transcription factors. *Genome Res* **22:** 1798–1812.

Whiteld TW, Wang J, Collins PJ, Partridge EC, Aldred SF, Trinklein ND, Myers RM, Weng Z. 2012. Functional analysis of transcription factor binding sites in human promoters. *Genome Biol* **13:** R50.

Wilczynski B, Furlong EE. 2010. Challenges for modeling global gene regulatory networks during development: Insights from *Drosophila*. *Dev Biol* **340:** 161–169.

Wunderlich Z, Bragdon MD, Eckenrode KB, Lydiard-Martin T, Pearl-Waserman S, Depace AH. 2012. Dissecting sources of quantitative gene expression pattern divergence between *Drosophila* species. *Mol Syst Biol* **8:** 604.

Yuh CH, Bolouri H, Davidson EH. 1998. Genomic cis-regulatory logic: Experimental and computational analysis of a sea urchin gene. *Science* **279:** 1896–1902.

Yuh CH, Bolouri H, Davidson EH. 2001. *Cis*-regulatory logic in the endo16 gene: Switching from a specification to a differentiation mode of control. *Development* **128**: 617–629.

Zuckerkandl E, Pauling L. 1965. Molecules as documents of evolutionary history. *J Theor Biol* **8**: 357–366.

FURTHER READING

The following books, reviews, and journal articles that are not specifically cited in this chapter but provide additional valuable information.

Alon U. 2006. An introduction to systems biology: Design principles of biological circuits. Chapman & Hall/CRC Mathematical and Computational Biology, Boca Raton, FL.

Babu MM, Luscombe NM, Aravind L, Gerstein M, Teichmann SA. 2004. Structure and evolution of transcriptional regulatory networks. *Curr Opin Struct Biol.* **14**: 283–291.

Bonn S, Furlong EE. 2008. *cis*-Regulatory networks during development: A view of *Drosophila*. *Curr Opin Genet Dev* **18**: 513–520.

Bonn S, Zinzen RP, Girardot C, Gustafson EH, Perez-Gonzalez A, Delhomme N, Ghavi-Helm Y, Wilczyński B, Riddell A, Furlong EE. 2012. Tissue-specific analysis of chromatin state identifies temporal signatures of enhancer activity during embryonic development. *Nat Genet* **44**: 148–156.

Borok MJ, Tran DA, Ho MC, Drewell RA. 2010. Dissecting the regulatory switches of development: Lessons from enhancer evolution in *Drosophila*. *Development* **137**: 5–13.

Bradley RK, Li XY, Trapnell C, Davidson S, Pachter L, Chu HC, Tonkin LA, Biggin MD, Eisen MB. 2010. Binding site turnover produces pervasive quantitative changes in transcription factor binding between closely related *Drosophila* species. *PLoS Biol* **8**: e1000343.

Carroll SB. 2008. Evo-devo and an expanding evolutionary synthesis: A genetic theory of morphological evolution. *Cell* **134**: 25–36.

Carroll SB, Prud'Homme B, Gompel N. 2008. Regulating evolution. *Sci Am* **298**: 60–67.

Carroll SB, Grenier JK, Weatherbee SD. 2005. From DNA to diversity: Molecular genetics and the evolution of animal design. Blackwell Scientific, Malden, MA.

Chorley BN, Wang X, Campbell MR, Pittman GS, Noureddine MA, Bell DA. 2008. Discovery and verification of functional single nucleotide polymorphisms in regulatory genomic regions: Current and developing technologies. *Mutat. Res.* **659**: 147–157.

Davidson EH. 2006. The regulatory genome: Gene regulatory networks in development and evolution. Academic Press, Burlington, MA.

Davidson EH, Levine MS. 2008. Properties of developmental gene regulatory networks. *Proc Natl Acad Sci* **105**: 20063–20066.

Ernst J, Kheradpour P, Mikkelsen TS, Shoresh N, Ward LD, Epstein CB, Zhang X, Wang L, Issner R, Coyne M, et al. 2011. Mapping and analysis of chromatin state dynamics in nine human cell types. *Nature* **473**: 43–49.

Furlong EEM. 2008. A topographical map of spatiotemporal patterns of gene expression. *Dev Cell* **14**: 639–640.

Gompel N, Prud'Homme B, Wittkopp PJ, Kassner VA, Carroll SB. 2005. Chance caught on the wing: cis-regulatory evolution and the origin of pigment patterns in *Drosophila*. *Nature* **433**: 481–487.

Gordon KL, Ruvinsky I. 2012. Tempo and mode in evolution of transcriptional regulation. *PLoS Genet* **8**: e1002432.

Hasty J, Isaacs F, Dolnik M, McMillen D, Collins JJ. 2001. Designer gene networks: Towards fundamental cellular control. *Chaos* **11:** 207–220.

Hasty J, McMillen D, Collins JJ. 2002. Engineered gene circuits. *Nature* **420:** 224–230.

Heintzman ND, Ren B. 2009. Finding distal regulatory elements in the human genome. *Curr Opin Genet Dev* **19:** 541–549.

Heintzman ND, Hon GC, Hawkins RD, Kheradpour P, Stark A, Harp LF, Ye Z, Lee LK, Stuart RK, Ching CW, et al. 2009. Histone modifications at human enhancers reflect global cell-type-specific gene expression. *Nature* **459:** 108–112.

Jeong S, Rebeiz M, Andolfatto P, Werner T, True J, Carroll SB. 2008. The evolution of gene regulation underlies a morphological difference between two *Drosophila* sister species. *Cell* **132:** 783–793.

Levine M. 2008. A systems view of *Drosophila* segmentation. *Genome Biol* **9:** 207.

Levine M, Davidson EH. 2005. Gene regulatory networks for development. *Proc Natl Acad Sci* **102:** 4936–4942.

Ong CT, Corces VG. 2011. Enhancer function: New insights into the regulation of tissue-specific gene expression. *Nat Rev Genet* **12:** 283–293.

Ong CT, Corces VG. 2012. Enhancers: Emerging roles in cell fate specification. *EMBO Rep* **13:** 423–430.

Savageau MA. 1976. Biochemical systems analysis: A study of function and design in molecular biology. Addison-Wesley, Reading, MA.

Spivakov M, Akhtar J, Kheradpour P, Beal K, Girardot C, Koscielny G, Herrero J, Kellis M, Furlong EE, Birney E. 2012. Analysis of variation at transcription factor binding sites in *Drosophila* and humans. *Genome Biol* **13:** R49.

Spitz F, Furlong EE. 2012. Transcription factors: From enhancer binding to developmental control. *Nat Rev Genet* **13:** 613–626.

Weirauch MT, Hughes TR. 2010. Dramatic changes in transcription factor binding over evolutionary time. *Genome Biol* **11:** 122.

Wilczynski B, Furlong EE. 2010. Dynamic CRM occupancy reflects a temporal map of developmental progression. *Mol Syst Biol* **6:** 383.

Visel A, Rubin EM, Pennacchio LA. 2009. Genomic views of distant-acting enhancers. *Nature* **461:** 199–205.

Zinzen RP, Girardot C, Gagneur J, Braun M, Furlong EE. 2009. Combinatorial binding predicts spatiotemporal *cis*-regulatory activity. *Nature* **462:** 65–70.

ONLINE RESOURCES

Systems Biology and Synthetic Biology

http://www.systemscenters.org NIH-supported National Centers for Systems Biology.

http://biobricks.org BioBricks Foundation.

http://www.biotapestry.org Biotapestry software for regulatory network visualization and modeling.

http://www.cytoscape.org Cytoscape graphical models organization.

http://igem.org International Genetically Engineered Machine (iGEM) Foundation.

http://partsregistry.org Registry of Standard Biological Parts.

http://sbml.org Systems Biology Markup Language organization.

ENCODE-Related Sites and Human Genetic Variation

http://cgap.nci.nih.gov Cancer Genome Anatomy Project.

http://encodeproject.org/ENCODE NIH ENCODE site.

http://factorbook.org TF binding data from ENCODE ChIP-seq.

https://www.gwascentral.org Database of GWAS studies.

http://www.modencode.org modENCODE project

http://www.mouseencode.org mouse ENCODE project.

http://www.nature.com/encode *Nature* link to ENCODE papers.

http://www.1000genomes.org 1000 Genomes Project, a deep catalog of human genetic variation.

http://regulomedb.org Database of SNPs in known and predicted regulatory regions.

Microbiology Resources

http://dbtbs.hgc.jp Database of transcriptional regulation in *Bacillus subtilis*.

http://ecocyc.org Ecocyc database of *E. coli*.

http://www.hmpdacc.org; http://commonfund.nih.gov/hmp Human Microbiome Project web sites.

http://www.microbesonline.org Microbes online database.

http://regtransbase.lbl.gov Bacterial regulatory factor and binding site database.

http://regulondb.ccg.unam.mx RegulonDB database of *E. coli* TFs and binding sites.

Databases of Transcription Factors, Binding Sites, and PWMs

http://www.cisreg.ca/cgi-bin/tfe/home.pl Transcription factor encyclopedia of TFs for human, mouse and rat.

http://www.gene-regulation.com/pub/databases.html TRANSFAC database of transcription factor binding motifs.

http://jaspar.cgb.ki.se/ JASPAR database of transcription factor binding motifs.

http://thebrain.bwh.harvard.edu/uniprobe UNIPROBE database of binding motifs from PBM data.

Organism-Specific Databases

http://bdtnp.lbl.gov/Fly-Net Berkeley *Drosophila* Transcription Network Project.

http://flybase.org Flybase for *Drosophila* and related species.

http://flyex.uchicago.edu/flyex Database of segmentation gene expression in *Drosophila*.

http://www.spbase.org/SpBase Sea urchin database.

http://www.wormbase.org Worm base for *C. elegans* and related species.

http://www.yeastgenome.org Yeast genome database.

http://www.yeastract.com; http://yetfasco.ccbr.utoronto.ca; http://stormo.wustl.edu/ ScerTF Yeast TF binding-site databases.

Index

www.ingramcontent.com/pod-product-compliance
Lightning Source LLC
Chambersburg PA
CBHW040147200326
41520CB00028B/7524